THE ROLE OF A SYLLABUS IN FLYING TRAINING

Most states have an aviation authority responsible for, among other matters, deciding which subjects must be understood by those training for a pilots' licence. Over the past quarter of a century the range of these subjects has grown considerably so that learning to fly, even when the student has no intention of becoming a professional pilot, is no longer as simple a matter as it was prior to the 1939–1945 war.

To ensure that all subjects and items within those subjects are taught during the course the value of a syllabus of training has in recent years assumed growing importance. It may be regarded as a shopping list of knowledge to be acquired but it has more to offer than that. A syllabus of flying training ensures:

(a) that the concept of the course is understood by the student and the instructor.
(b) that the content is known by all concerned with the course of training.
(c) that progress during the course may be monitored.
(d) that space is provided where the student may record the instructor's comments after each lesson. These are of value during revision and subsequent practice.
(e) that a summary of learning and practice is available to both student and instructor, thus ensuring that the course has been completed.

Flight Briefing for Pilots, Vols. 1, 2 and 3, have been written to conform with a Syllabus of Flying Training for the Private Pilots' Licence and the IMC Rating (currently confined to the UK) that has been recognised by the UK Civil Aviation Authority. The syllabus is published by Longman.

Throughout the text headings published in this type [**Instrument Indications**] cross refer to the syllabus.

GW00322359

Books on Aviation by N. H. Birch and A. E. Bramson

Flight Briefing for Pilots series
Flying the VOR
A Guide to Aircraft Ownership
The Tiger Moth Story
Captains and Kings
Radio Navigation for Pilots
Flight Briefing for Microlight Pilots
Flight Emergency Procedures for Pilots

By Alan Bramson

Be a Better Pilot
Make Better Landings
The Book of Flight Tests
Master Airman, a biography of
   Air Vice-Marshal Donald Bennett CB, CBE, DSO, FRAeS
Principles of Flight (Audio-visual trainer)

By Neville Birch

The Instrument Rating
Passenger Protection Technology in Aircraft Accident Fires

Flight Briefing for Pilots

Volume 2.
Ground Subjects for the Private Pilot's Licence.

This manual has been written to conform with the UK Civil Aviation Authority recognized Syllabus of Flying Training for the Private Pilot's Licence (Aeroplanes) which is published by Longman.

Flight Briefing for Pilots

Volume 2
Ground Subjects for the Private Pilot's Licence

N. H. Birch, PhD, MSc, MRAeS
Liveryman of the Guild of Air Pilots and Air Navigators
and A. E. Bramson, FRAeS
Former Chairman of the Panel of Examiners
Liveryman of the Guild of Air Pilots and Air Navigators
Illustrated by A. E. Bramson

FOURTH EDITION

Longman
Scientific &
Technical

Longman Scientific & Technical
Longman Group UK Limited,
Longman House, Burnt Mill, Harlow,
Essex CM20 2JE, England
and Associated Companies throughout the world.

First published in Great Britain
  by Pitman Publishing Limited 1968
Reprinted 1969, 1970, 1972
Second Edition 1974
Reprinted 1975, 1976, 1977
Third Edition 1979
Revised reprint 1984
Fourth Edition published
  by Longman Group Limited 1986
Second impression 1988

British Library Cataloguing in Publication Data

Birch, N. H.
Flight briefing for pilots. -4th ed.
Vol. 2: Ground subjects for the private pilot's
licence
1. Airplanes - Piloting
I. Title II. Bramson, A. E.
629.132′52   TL210

ISBN 0-582-98815-2

Library of Congress Cataloging-in-Publication Data

Birch, Neville Hamilton.
  Flight briefing for pilots.

  Vol. 2: 4th ed.
  Includes indexes.
  Contents: v. 1. An introductory manual for flying
training complete with air instruction ——v. 2.
Ground
subjects for the private pilots' licence.
  1. Airplanes——Piloting.   I. Bramson, Alan
Elesmere.
II. Title.
TL710.B542   1986   629.132′52   85-23296
ISBN 0-582-98814-4 (v. 1)
ISBN 0-582-98815-2 (v. 2)

Produced by Longman Singapore Publishers Pte Ltd
Printed in Singapore

Contents

Plates 1–8 (Clouds) are position between pp. 100 and 101.

This book has been written to conform with a UK Civil Aviation Authority recognised syllabus of training for the Private Pilot's licence published by Longman.

Preface

List of Abbreviations

The basis of successful flying instruction is vested in a thorough knowledge of all the related ground subjects. This second volume (previously *Vol. 4*) in the *Flight Briefing for Pilots* series should therefore be studied before the flying exercises have begun.

In effect this book is a companion to *Flight Briefing for Pilots, Vol. 1*, the Manual of Flying Training for the Private Pilot's Licence and, whenever possible, subjects relating to another chapter or volume are cross-referenced.

At the end of each chapter are typical multiple-choice questions, similar in style to those likely to form part of the Licence Examinations. The answers will be found in Appendix I.

Volume 2 is intended to provide the student with a basic knowledge of the ground subjects which will enable him or her to pass the licensing authority examinations for the Private Pilot's Licence and at the same time act as a first step towards a professional licence and/or a Flying Instructor's Rating.

Many books have been written on each of the subjects dealt with in this volume. However, most readers will be aiming to become pilots, not engine designers, airframe stressmen or specialists in aviation law. Consequently, we have confined the text to essentials and then applied ourselves to making the complicated appear simple. Since Volume 1 first appeared in 1961 it has been our constant aim to avoid turning the practical subject of airmanship into an academic exercise.

Since the *Flight Briefing for Pilots* series was first written flying has become more complicated with the introduction of advanced equipment, even in light, single-engine aircraft. However, the authors sympathize with those who are not 'mechanically minded' and they have endeavoured to write this manual with such readers in mind.

NHB
AEB

List of Abbreviations

AC	altocumulus
A/H	alter heading
a.m.s.l.	above mean sea level
AI	attitude indicator
AIP	Air Information Publication
AIS	Aeronautical Information Service
AS	altostratus
ASI	airspeed indicator
ATA	actual time of arrival
ATCC	Air Traffic Control Centres
BHP	brake horsepower
C	centigrade/celsius
C. of A.	Certificate of Airworthiness
C of G	centre of gravity
CB	cumulonimbus
CC	cirrocumulus
CI	cirrus
CS	cirrostratus
CSU	constant speed unit
CU	cumulus
Dev.	deviation
DI	direction indicator
DR	dead reckoning
EAT	expected approach time
ETA	estimated time of arrival
ETD	estimated time of departure
FIR	flight information region

FIS	flight information service
GPH	gallons per hour
G/S	ground speed
Hdg	heading
Hdg (C)	compass heading
Hdg (M)	magnetic heading
Hdg (T)	true heading
HP	horsepower
HPa	hectopascals
ht	height
IAS	indicated airspeed
ICAO	International Civil Aviation Organization
IFR	instrument flight rules
IMC	instrument meteorological conditions
ISA	International Standard Atmosphere
km	kilometre
kt	knot
lb	pound
L/D	lift/drag
MTWA	Maximum Total Weight Authorized
M	magnetic
m.b.	millibar
mph	miles per hour
m.s.l.	mean sea level
n.m.	nautical mile
Notams	notices to airmen
NS	nimbostratus
OAT	outside air temperature
P	port
p.p.	pin-point
posn.	position

RAS	rectified airspeed
RPM	revs. per minute
RTF	radiotelephony

S., stb.	starboard
SC	stratocumulus
S/H	set heading
ST	stratus
st. m.	statute mile

T	true (direction, speed)
TAF	Terminal Aerodrome Forecast
TAS	true airspeed
Tr. Req.	required track
TMG	track made good

var.	variation
VFR	visual flight rules
VMC	visual meterological conditions
VSI	vertical speed indicator

W/V	wind velocity

Chapter 1
Aviation Law

Those learning to drive a motor vehicle are confronted by a formidable catalogue of regulations, statutory and advisory. The Road Traffic Act and Highway Code are two such examples. Aviation has its own comparable regulations.

Like his motoring counterpart the student pilot will have an instructor who, among other tasks, will ensure that all flying training is conducted safely and in accordance with the law. Information related to procedures, practices and the law is in the main found in the *Air Information Publication* (AIP) (CAP 32) and the *Air Navigation Order*.

The *Air Navigation Order* is a legal document. As such it is not the easiest of publications to use, however, the AIP (*Air Pilot*) has been compiled to interpret it. Furthermore, *Aviation Law for the Private Pilot's Licence* (CAP 85) contains most of the relevant information in summary form and it will be necessary to study this publication before taking the written examination for the PPL.

From time to time the regulations are changed, enlarged or amended, therefore this chapter is confined to that information which is of immediate or early concern to the student pilot. It is the responsibility of all pilots to keep themselves informed of changes to the regulations.

The syllabus requirements for aviation law will be extended, they are listed on page 32.

Pilot's licences

There are four different classes of licence for air pilots and in order of seniority they are

Title	Abbreviation
Airline Transport Pilot's Licence	ATPL
Senior Commercial Pilot's Licence	SCPL
Commercial Pilot's Licence	CPL
Private Pilot's Licence	PPL

Requirements for the private pilot's licence

Flying Experience

A minimum of 43 hours as pilot to include not less than:

(*a*) 20 hours' dual instruction under the supervision of a qualified flying instructor

(*b*) 10 hours' as pilot in command

The above minimum experience must also include:

(*a*) 4 hours' instruction in pilot navigation

(*b*) 4 hours' solo cross-country flying

(*c*) 4 hours' instruction in instrument flying

(*d*) 2 hours' stall/spin awareness and avoidance training

Approved Course

Where an approved course is undertaken, namely at training establishments which meet specific Civil Aviation Authority (CAA) criteria, the course is reduced to a minimum of 38 hours' experience to include the above requirements and must be completed within a period of 6 months.

Age

The minimum age both for flying an aircraft solo and for the award of a Private Pilot's Licence is 17 years.

Medical

A medical examination conducted by an authorized medical examiner is necessary before flying solo. The medical certificate is valid for 2 years under the age of 40, thereafter for 1 year in addition to the remainder of the month of issue.

General Flight Test and Technical Examination

A General Flight Test (normally of 1½ hours' duration) in which the candidate will be required to demonstrate competence based on the air exercises detailed in Volume 1 of this series.

A Technical Examination in four subjects, consisting of:

(*a*) Aviationn Law, Flight Rules and Procedures
(*b*) Navigation and Meteorology
(*c*) Aeroplanes Technical Part I (General)
(*d*) Aeroplanes Technical Part II (Specific Aircraft Type), an oral examination.

The Private Pilot's Licence is valid permanently from the date of issue subject to:

(*a*) A current medical certificate;
(*b*) A current certificate of experience or test.

The licence will show the ratings to which the holder is entitled, e.g. night rating and the aircraft classification which is sub-divided into four groups:

Group A Single-engined aircraft below 5700 kg maximum take-off weight authorized;
Group B Multi-engined below 5700 kg MTWA;
Group C Aircraft over 5700 kg MTWA;
Group D Aircraft below 150 kg (microlights).

In order to maintain a group on a PPL every 13 months the holder must produce evidence of 5 hours' in-command flying experience on an aircraft in that group, his licence and log book being checked by an authorized examiner who, on being satisfied, will sign the certificate of experience (mentioned under heading (*b*). When a licence holder has not completed the qualifying 5 hours' flying to maintain a group, retention of that group will be subject to a satisfactory flying test conducted by an examiner who will sign the certificate of test.

A PPL entitles the holder to fly passengers, but not for hire or reward in weather conditions related to visual flight (p. 9).

Ratings

As knowledge and flying experience increase the private pilot may obtain the following ratings:

1. Aircraft Ratings
2. Night Rating
3. Flight Radiotelephony Licence
4. Flying Instructor's Ratings
5. The IMC Rating
6. Instrument Rating.

1. Aircraft Ratings

As already explained the PPL entitles the holder to fly aircraft in any or all of four groups of aircraft. Invariably the licence is first issued with light single-engined aircraft (Group A) privilege and a short conversion course and technical examination will be necessary before the pilot can attempt the flying test for aircraft in other groups. For aircraft weighing more than 5700 kg a Civil Aviation Authority specific type technical examination is required.

2. Night Rating

No pilot may fly passengers at night unless he has a night rating and has completed 5 take-offs and landings in the preceding six months. The requirements for this addition to the PPL are:

1. A minimum of 50 hours' flying of which half must be as a pilot in command;
2. A minimum of 5 hours' instruction on instrument flying of which half may have been practised in a simulator;
3. A minimum of 5 hours' night flying under the supervision of a qualified instructor in which must be included at least five separate flights as pilot-in-command, this to be completed within six months of making application for the rating.

3. Flight Radiotelephony Licence

With the development of all types of flying activity it is necessary for all pilots to be able to communicate with the various air traffic services on the ground. The rapid increase in air traffic has brought with it a similar growth in profusion of radio communications so that brevity in passing information and a knowledge of correct procedures is essential if smooth working of the Air Traffic Services is to be ensured. The Licence is gained by a practical test.

4. Flying Instructor's Ratings

Whereas any motorist may teach another person to drive a car, flying instruction by pilots not qualified as instructors is not allowed.

There are two classes of Flying Instructor's Ratings. The requirements are as follows:

1. Assistant Instructor's Rating

(i) Pass the pre-entry examination.

(ii) A valid licence.

(iii) A minimum of 150 hours as pilot-in-command, and at least 30 hours' single-engine experience and 35 hours' dual instruction on the class of aircraft to which the application relates.

(iv) An approved flying instructor's course of not less than 25 hours and 50 hours' ground instruction in teaching methods.

(v) An IMC or Instrument Rating.

(vi) Pass a test by an authorized examiner.

Holders of the Assistant Flying Instructor's Rating may only instruct under the supervision of an instructor with the more senior rating now described. The Assistant Flying Instructor's Rating must be renewed every 13 months.

2. Flying Instructor's Rating

Candidates for this rating must produce evidence of:

(i) A valid pilot's licence.

(ii) A minimum of 400 hours as pilot-in-command.

(iii) A minimum of 200 instructional hours and 6 months as an Assistant Instructor or as a qualified military instructor.

(iv) A valid IMC or Instrument Rating.

Both ratings are confined to day instruction unless the applicant has completed 10 hours' flying by night of which up to 5 hours may have been dual. Two hours of the dual night flying must have been flown in the seat normally occupied by the flying instructor and in the company of an approved instructor. The night flying instructor course also includes 5 hours' ground instruction. There are separate ratings for aerobatics, instrument flying and multi-engine training. The Flying Instructor's Rating is renewed every 25 months.

At the few schools training airline pilots, instructors are tested by Civil Aviation Authority examiners, all other instructors and instructor candidates being tested by the Panel of Examiners, a small committee of very experienced flying instructors granted delegated authority for the purpose by the Civil Aviation Authority.

5. Instrument Rating

Flying an aircraft over open country on a clear day is within the capabilities of quite inexperienced pilots. Flying an aircraft through areas of intense traffic under conditions of low cloud and poor visibility is quite a different matter. It will shortly be explained that certain areas of high activity (e.g. the London, New York or Paris areas) are of necessity subject to rigid control by the air traffic services. It is imperative that pilots flying within **Controlled Airspace** should be capable of maintaining both height and position to very high standards of accuracy, using the flight instruments and radio aids provided in the aircraft. Only pilots with an instrument qualification may fly within controlled airspace and it therefore follows that this is an essential qualification for all pilots engaged on air transport, and a great asset to the non-professional pilot using an aircraft for business purposes.

The principal requirements for an Instrument Rating are:

(i) Completion of an instrument flying course at a CAA approved flying school;
(ii) At least 150 hours' pilot in command experience of which 50 hours must be cross-country flying;
(iii) At least 40 hours' instrument flying of which 20 hours may have been conducted on a simulator.

There is an examination and a stringent flying test which is conducted by a CAA examiner. This is explained in detail in Volume 6 of the new series *Flight Briefing for Pilots*.

6. The IMC Rating (valid in the UK only)

The IMC Rating is an instrument qualification which a pilot may obtain to extend the privileges of his licence, permitting the holder to fly under these conditions:

Outside controlled airspace (without passengers)
No limitations

Outside controlled airspace (with passengers):
(i) Above 3000 ft a.m.s.l. in IMC conditions;
(ii) Below 3000 ft with a visibility of 3 n.m.

Within controlled airspace, with a Special VFR Clearance:
visibility is not less than $1\frac{1}{2}$ n.m.
1 n.m. when taking off or landing.

Requirements

Candidates for the rating must have at least 100 hours' pilot experience to include 60 as pilot-in-command, of which 30 hours are cross-country flights. Applicants must hold an R/T licence, and have completed an approved course of 10 hours' instrument flying instruction.

Test

During the air test candidates must fly on instruments within certain limits of accuracy. There is an oral examination on the following subjects: flight planning; en-route and approach navigation using VDF and radar; meteorology; altimeter setting procedures; use of the AIP and other publications for the planned use of radio facilities; and the privileges and limitations of the IMC rating itself.

A full explanation of the IMC Rating is contained in Volume 3 of the new series *Flight Briefing for Pilots*. Currently the rating is only valid within the UK.

Rules for avoiding collision

In the air
1. Aeroplanes must give way to airships, gliders and balloons.
2. When two aircraft are approaching head on each shall alter course to the right (Fig. 1).
3. When two aircraft are converging the one which has the other on its right shall give way (Fig. 2).
4. An aircraft overtaking another shall keep to the right and remain clear of that aircraft whatever its relative position (Fig. 3).

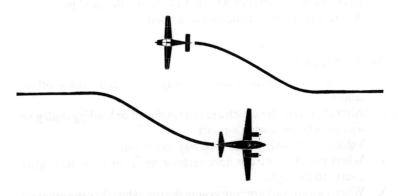

Fig. 1   Approaching head-on.

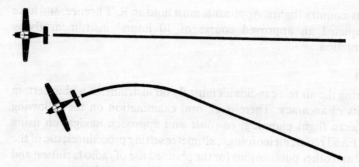

Fig. 2   Converging aircraft.

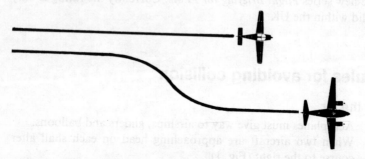

Fig. 3   Overtaking another aircraft.

5.   An aircraft approaching to land has right of way over aircraft in flight or on the ground.
6.   When two or more aircraft are approaching to land the lower has right of way except where ATC have allocated priorities. Aircraft in an emergency have priority.

On the ground

1.   Aircraft and vehicles shall give way to aircraft taking off or landing.
2.   Aircraft or vehicles which are not taking off or landing shall give way to vehicles towing aircraft.
3.   Vehicles not towing shall give way to aircraft.
4.   When two aircraft are approaching head on each shall alter course to the right.
5.   When two aircraft are converging the one which has the other on its right shall give way.

6.   An aircraft which is being overtaken has right of way and the overtaking aircraft is responsible for safe clearance.

·   *Note*: Except when the seats are arranged in tandem the pilot in command is placed on the left side of the aircraft unless the pilot is a student under training with a qualified instructor or check pilot. To comply with these rules this seat must be occupied.

Air Traffic Control

With the increase in air traffic of all categories has come the need to ensure adequate separation of aircraft, particularly over important centres of communication. This is intensified under conditions of low cloud and/or poor visibility and a little thought will reveal that whereas in good weather a pilot may be well able to integrate with other traffic, poor visibility will prevent him from seeing other aircraft or he may lose sight of the ground (or both). Control and separation of aircraft is placed in the hands of a ground organization known as the **Air Traffic Service** and flights within the UK will be conducted in accordance with the following rules:

Visual Flight Rules   (VFR)
Special Visual Flight Rules   (SVFR) or
Instrument Flight Rules   (IFR)

Visual Flight Rules

	Flight Visibility (n.m.)	Distance from Cloud
In Controlled Airspace	5	1 n.m. horizontally 1000 ft vertically
Outside Controlled Airspace Above 3000 ft a.m.s.l.	5	1 n.m. horizontally 1000 ft vertically
At or below 3000 ft a.m.s.l.	3	1 n.m. horizontally 1000 ft vertically
At or below 3000 ft a.m.s.l. at airspeed 140 kt or less	1	Clear of cloud and in sight of the ground

Should any of these conditions deteriorate Instrument Meteorological Conditions (IMC) will prevail then the flight should be conducted in accordance with IFR. The following observations will help the student to understand the division of responsibility between pilots and the Air Traffic Service:

(i) When flying under VFR safe conduct of the flight and separation from other aircraft are the responsibility of the pilot.

(ii) Irrespective of weather conditions permanent Instrument Flight Rules apply to certain Controlled Airspace (Rule 21).

(iii) In the United Kingdom flight at night irrespective of weather conditions or type of airspace is deemed to be under IFR.

(iv) Non-Instrument-Rated pilots wishing to fly under the conditions noted in paragraphs (ii) and (iii) must obtain a **Special VFR Clearance** which may be granted by the Air Traffic Service responsible.

(v) A Special VFR Clearance allows the aircraft to fly in controlled airspace, clear of cloud and in sight of the ground complying with ATC instructions.

(vi) When Instrument Meteorological Conditions prevail pilots wishing to enter Controlled Airspace must obtain clearance from the appropriate Air Traffic Control (ATC) unit. Clearance will only be granted when:

   (*a*) the pilot has an Instrument qualification and the aircraft is equipped with the appropriate radio communications and navigational equipment or

   (*b*) a Special VFR Clearance has been obtained and/or

   (*c*) traffic and meteorological conditions permit.

Within controlled airspace separation of aircraft flying under IFR is the responsibility of the Air Traffic Service.

Controlled Airspace

Organization and procedure as it relates to air traffic control is to a considerable extent international. In so far as the UK is concerned England, Scotland, Wales and Northern Ireland are divided into two **Flight Information Regions (FIR)** which extend from ground to Flight Level 245. Above this level are the **Upper Flight Information Regions (UIR)**. The two FIR/UIRs are known as 'Scottish', and 'London', and lie north and south of a line drawn east–west approximately through Newcastle upon Tyne. Each FIR/UIR is controlled by an **Air**

Traffic Control Centre (ATCC) which provides the following services within the confines of its area:
1. a flight information and alerting service;
2. an air traffic advisory service to aircraft flying under IFR within advisory areas and routes;
3. control of aircraft flying under IFR within the **Airways** situated in the FIR. The various services are available on different radio frequencies so that a number of pilots may simultaneously call the required ATS. The FIR/UIR is not controlled airspace although it includes a system of Zones/Areas and Airways within it which are controlled.

Control Zones

As already explained certain major airports are subjected to heavy traffic and it follows that flights must be integrated. Approaches to some major airports are surrounded by a **Zone** which extends upwards from ground level and where permanent Instrument Flight Rules are adopted regardless of weather (Rule 21). Nevertheless in common with controlled airspace generally a pilot may request clearance into or through this airspace. In such circumstances aircraft equipped with **Secondary Surveillance Radar (SSR)** i.e. a **Transponder** may well be at an advantage.

Control Area

Similar to a Control Zone but extending upwards from a specified height above ground level to an upper defined flight level. A **Terminal Control Area** is part of a control area where a number of airways converge, typical examples occurring over the London, Manchester and Edinburgh areas.

Airways

Linking the Control Zones and in effect providing protected corridors between centres of heavy traffic are the *Airways*. Of 10 nautical miles' width the Airways extend vertically from a base of 3000 ft or more to a designated upper level. Aircraft wishing to cross an airway without entering may fly below its base (terrain clearance permitting) but care should be taken to determine the base of the airway at the point of crossing since the lower limit may be stepped down as the airway approaches a junction with a control zone (Fig. 4).

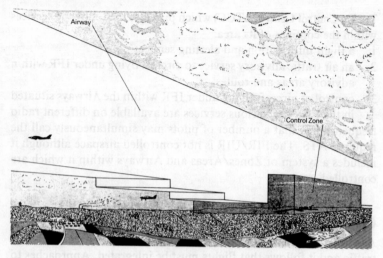

Fig. 4   An airway, its base stepping down as it joins a control zone.

The Control Zones and their linking Airways are illustrated in Fig. 5, details being shown on the relevant aviation charts and illustrated in the RAC Section of the AIP.

Advisory Service Area

In certain parts of the FIR **Advisory Service Areas** exist and some of this airspace is in the form of extended corridors which are known as **Advisory Routes**. Although pilots flying outside controlled airspace are responsible for their own separation under IFR, when using this service separation is co-ordinated by ATC.

Special Rules Area/Zone

These areas have defined upper and lower limits in which aircraft are required to comply with ATC instructions in IMC or VMC conditions. Similar to the **Special Rules Area** is a **Special Rules Zone** except the base of the airspace is at ground level.

Military Airfield Traffic Zones

At specified military airfields traffic zones are designated within a 5 mile radius from ground level to 3000 ft and a projected airspace 4 miles wide along the extended runway centre line. The zone,

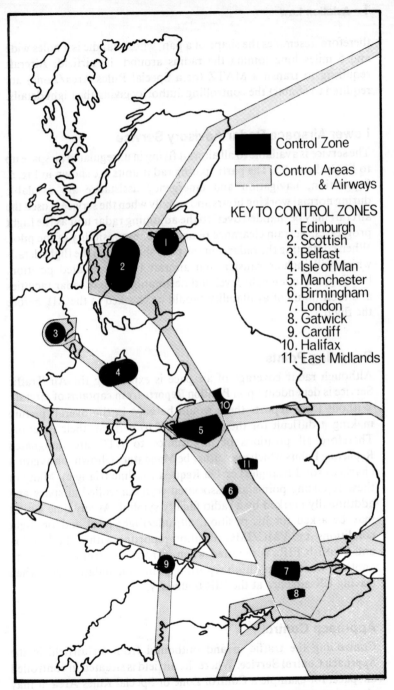

Key to control zones:
- Control Zone
- Control Areas & Airways

KEY TO CONTROL ZONES
1. Edinburgh
2. Scottish
3. Belfast
4. Isle of Man
5. Manchester
6. Birmingham
7. London
8. Gatwick
9. Cardiff
10. Halifax
11. East Midlands

Fig. 5   Simplified version of the UK airways and controlled airspace system.

therefore, resembles the shape of a pan, whose handle is 4 miles wide and 5 miles long joining the radius around the airfield. Aircraft requiring to transit a MATZ (or a Special Rules Area/Zone) are required to contact the controlling authority giving the flight details.

Lower Airspace Radar Advisory Service

The service is available to all aircraft flying in unregulated airspace up to Flight Level 95. The participating radar units are shown in Fig. 6. Separation, navigation and emergency assistance are available during normal working hours and usually when the service is used the aircraft may be 'handed over' to the adjoining radar unit as the flight progresses. Terrain clearance remains the responsibility of the pilot. When contacting the radar unit the pilot should pass the flight details which will include callsign and aircraft type, estimated position, heading, altitude/flight level and destination. These radar facilities and their hours of availability should be checked in the AIP before the flight.

Reporting Points

Although radar coverage of airspace is expanding the Air Traffic Service is dependent upon **Position Reports** from captains of aircraft. Without these reports there would be gaps in the overall picture making it difficult for the control staff to perform their function. Therefore all positions of importance to ATC are designated **Reporting Points** which may either be **Mandatory** (shown on aviation charts as solid triangles) or **On Request** (outline triangle). Many of these reporting points are associated with controlled airspace and additionally marked by a radio facility symbol. At any time a pilot may be asked for his position, altitude/flight level and/or flight conditions, i.e. VMC/IMC. Position reporting within the London and Scottish FIRs is required when transferring from one FIR to another, when the position reporting point is mandatory and when leaving UK airspace at the FIR boundary.

Approach Control

Controlling the traffic in and outbound from an airfield is the **Approach Control Service**. Where the airfield is situated in controlled airspace, for example a Control Zone or Special Rules Area, it may also control that airspace and aircraft in transit through it. Approach

LOWER AIRSPACE RADAR ADVISORY SERVICE

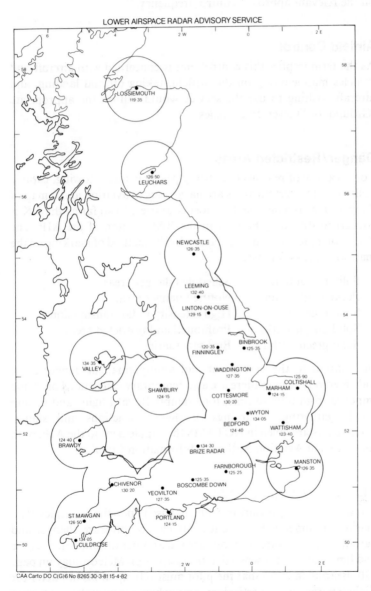

Fig. 6   Lower Airspace Radar Advisory Service.

Control will usually be supported with radar. VDF and airfield NDB(s). Aircraft wishing to contract Approach Control would do so on the relevant approach control frequency.

Airfield Control

As the term implies this control unit is concerned with aircraft and vehicles manoeuvring on the airfield, taking off and landing, and aircraft wishing to use the service would call on the appropriate 'Ground' or 'Tower' frequencies.

Danger/Restricted Areas

For a variety of reasons (e.g. firing practice, target towing, captive balloons, etc.) certain areas are hazardous to aviation and a map of danger areas is available for most countries. In so far as the UK is concerned, this may be seen in the RAC section of the AIP. The degree of danger may be recognized by the method of marking on the map which is as follows:

1. Solid red outline   Scheduled danger areas.
2. Pecked red outline   Notified danger area.
3. Solid blue outline   High intensity radio transmissions.
4. Solid purple outline   Prohibited and restricted areas.
5. Solid green outline   Bird sanctuaries.

In addition to the permanent and semi-permanent danger areas indicated on the map there are also temporary restrictions on flying imposed from time to time, for example Royal Flights and certain public gatherings. Warnings of these restrictions and similar occurrences are given in NOTAMS the supplement to the AIP, or in the UK Civil Aviation Authority Information Circulars.

Altimeter Setting Procedure

The altimeter is a pressure instrument based upon the principles of an aneroid barometer. Because it is affected by pressure changes and variations in pressure from one area to another the altimeter, its use and limitations are discussed in the chapter on Meteorology on page 120. Bearing in mind that the pilot must rely upon his altimeter for flight separation (i.e. vertical separation between aircraft) and terrain clearance, the importance of the instrument and the dangers inherent in its misuse should be fully understood.

Flights Outside Controlled Airspace and the Quadrantal Rule

When flying outside controlled airspace or advisory areas pilots must assume responsibility for their separation from other aircraft. Although air traffic outside controlled airspace is of relatively low density the risk of collision with other aircraft is always a possibility and vigilance is the maxim.

The general flight rules and rights of way have already been explained on page 7 and these provide a degree of flight separation.

Pilots flying in IMC above transition altitude must set their altimeters to the standard setting of 1013.2 mb, thus ensuring that all aircraft are flown with altimeters using the same datum.

Cruising levels are chosen according to the aircraft's magnetic track at the time. At first examination it would appear simpler to use compass headings until it is realized that a strong cross-wind will subject a 90 kt light aircraft to considerable drift whereas a 500 kt jet aircraft on the same compass heading would be little affected. Although both aircraft may be steering the same compass heading they could well be on converging tracks with the attendant risk of collision. Therefore pilots flying outside controlled airspace and above 3000 ft will achieve separation by:

(a) determining their magnetic track and

(b) choosing a cruising level appropriate to that track, using the **Quadrantal Rule**, which in effect divides the compass into four sectors and confines aircraft to certain levels within these sectors (altimeter set to 1013.2 mb).

The quadrantal rule may easily be remembered and is arranged as follows:

Flights between 3000 ft and below 24,500 ft	
Magnetic Track	*Cruising Level*
000° to 089° inclusive	Odd thousands of feet
090° to 179°	Odd thousands of feet + 500 feet
180° to 269° inclusive	Even thousands of feet
270° to 359° inclusive	Even thousands of feet + 500 feet

The quadrantal rule as it applies up to Flight Level 245 and the separation achieve are illustrated in Fig. 7.

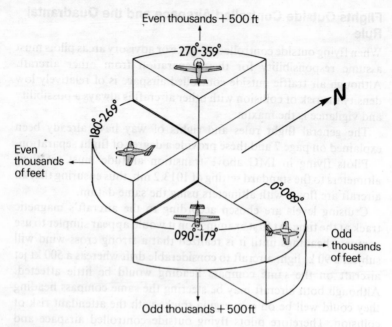

Fig. 7 The "Quadrantial Rule" as it applies up to Flight Level 245.

Minimum Safe Altitude (Low Flying)

The exhilaration of low flying has a fascination which for some pilots may be likened to the moth and candle. Although there are certain dangers inherent in low flying it is nevertheless an exercise to be practised under proper conditions and correct supervision while observing the regulations. The pilot who understands the principles of low flying is less likely to find himself in difficulty should an emergency occur which can only be resolved by continuing the flight below lowering cloud and in sight of the ground.

Before it is attempted solo, low flying must be carried out under the guidance of a qualified flying instructor. Selected low flying areas are sometimes available for this exercise (explained in Chapter 16 Volume 1) but in any event it should be practised over open country free from power cables, television masts, smoke stacks or other obstructions and never near hospitals, houses or farm animals. The law relating to the minimum safe altitude of aircraft is as follows:

(*a*) An aircraft other than a helicopter shall not fly over any congested area of a city, town or settlement below:

    (i) such height as would enable the aircraft to alight clear of the area and without danger to persons or property on the surface in the event of failure of a power unit, or

    (ii) a height of 1500 ft above the highest fixed object within a distance of 2000 ft.

(*b*) An aircraft shall not fly:

    (i) over or within 1000 yards of any assembly in the open air of more than 1000 persons assembled for the purpose of witnessing or participating in any organized event, except with the permission in writing of the CAA and in accordance with any conditions therein specified and with the consent in writing of the organizers of the event; or

    (ii) below such a height as would enable it to alight clear of the assembly in the event of the failure of a power unit.

    (*Note*. If a pilot charged with contravention of this subparagraph is able to show that he flew within the limits mentioned for reasons not connected with the assembly or event, this proof will be regarded as a good defence.)

(*c*) An aircraft shall not fly closer than 500 ft to any persons, vessel, vehicle or structure.

Exceptions. The foregoing shall not apply during:

(*a*) Organized air racing

(*b*) Organized flying displays

(*c*) Taking off and landing or practice approaches to landings when confined to airspace normally used for the purpose

(*d*) Flying to check navigational aids or procedures at Government or licensed airfields

(*e*) Flying as may be necessary to save life

(*f*) Flight under Special VFR in accordance with ATC instructions.

Signals

Although most aircraft carry radio, many of the smaller airfields are without radio facilities so that some means of conveying airfield and circuit information must be provided for pilots arriving over the landing area and those on the ground who are about to depart. These functions are performed by the **Signals Area**, a 40 ft x 40 ft hollow white square situated near the control tower. Within the signals area are displayed various signals which are of sufficient size and prominence to be easily seen from the air by pilots crossing the airfield at a height of 2000 ft or so. The signals area includes a mast on which are hung various signals for the benefit of pilots on the ground. Since the advent of aircraft radio these signals are of less importance than hitherto but they must nevertheless be understood by pilots for the reasons given at the beginning of this section. The various signals are illustrated in Appendix II at the end of the book.

Aircraft Navigation Lights

Aircraft cleared for night flying must be equipped with lights as follows:

Red light on PORT (left) wingtip.
Green light on STARBOARD (right) wingtip.
White light on TAIL of aircraft.

The lights are designed so that only one colour may be seen at any angle and it is the practice with some light aircraft to carry a three sector light above the fuselage and a similar one below. Angle of vision for the three lights are illustrated in Fig. 8.

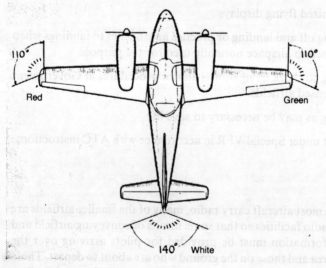

Fig. 8   Aircraft navigation lights.

Anti-collision Beacons and Strobes

To assist collision avoidance by making aircraft more conspicuous red flashing **Anti-collision Beacons** are usually fitted. Sometimes these are supplemented by wingtip-mounted **Strobe Lights**, white high-intensity lights which flash in a similar manner to those used in photography. Beacons and strobes are used by day and by night but strobes can be distracting while flying in cloud.

Identification Beacons

The larger civil and military airfields have a beacon which is arranged to flash the airfield identification in a two-letter Morse signal every 12 seconds (green light at civil and red light at military airfields). Alternatively, Airlight Beacons flashing a rotating green/white beam may be found at some civil airfields.

Visual signals

The following visual signals are used by Airfield Control to regulate the movements of aircraft.

To an aircraft *on the ground*

Intermittent green beam	Taxi within the manoeuvring area
Steady green beam	Take off
Intermittent red beam	Taxi clear of landing path immediately
Steady red beam	Remain stationary within manoeuvring area
Intermittent white beam	Return to parking position

To an aircraft *in flight near an airfield*

Intermittent green beam	Return to the airfield and wait for permission to land
Steady green beam or green Very signal	Permission to land
Intermittent red beam	Permission to land cannot be given; land at another airfield
Steady red beam or a red Very signal	Landing temporarily suspended.

To an aircraft *in flight*

A number of black or white smoke-puff projectiles, or white stars bursting at intervals, or an intermittent white beam flashed in the direction of the aircraft	A prohibited area is being approached; after heading immediately
A series of bursting green stars	A prohibited area has been violated; land at nearest airfield

Safety measures

Imperilling safety

A person may not act in a way that may endanger the safety of an aircraft or fly an aircraft in a manner that may endanger persons or property. No one may enter an aircraft when drunk or be drunk in an aircraft nor shall anyone acting as pilot or crew be affected by drink or drugs.

The carriage by aircraft of the following is prohibited:

(a) War munitions, including guns, grenades, explosives, etc.
(b) Dangerous goods, e.g. toxic substances.

Fuel

Petrol and oil must be carried in the approved stowages.

Smoking

Notices permitting or prohibiting smoking must be displayed in all aircraft.

Aerobatics

Aerobatics must not be carried out unless:

(a) The aircraft has been certificated for aerobatics.
(b) The pilot is competent to perform the manoeuvres.
(c) The aircraft is over open country and clear of Controlled Airspace. Within controlled airspace permission of ATC is required.

Dropping of objects

Without the permission of the CAA no object shall be dropped from an aircraft except (i) for the purpose of saving life, (ii) jettisoning fuel or articles in an emergency, (iii) ballast of fine sand and water, (iv) tow ropes provided there is no danger to persons or property.

Action Before Flight – pilot's responsibilities

Before taking-off a pilot must ensure that:

(a) The aircraft he is about to fly has been inspected and that the certificate of maintenance is valid.

(b) The equipment and instruments necessary for the flight are serviceable and safety devices are correctly stowed in the aircraft.

(c) The aircraft is properly loaded within the permitted C. of G. range.

(d) Fuel and oil are sufficient for the flight.

(e) The wings, tail-plane and control surfaces are free from snow, ice and frost.

(f) The flight can be conducted safely with regard to the weather (obtain relevant met. information), performance of the aircraft, terrain clearance to the destination and/or the alternative airfield.

Note. Anything which is found to be unserviceable in the aircraft during the flight must be reported on landing.

Aircraft accidents

If an aircraft is seriously damaged in an accident, or a person involved in an aircraft accident is killed or injured, it is the duty of the pilot, or operator concerned to report the details to the Accident Investigation Branch (Dep. of Transport). Details should be sent as soon as possible after the accident and the local police should also be informed.

If the accident occurs outside the British Isles a notice in writing is required by the Civil Aviation Authority.

Removal of aircraft

An aircraft which is seriously damaged in an accident, e.g. has fuselage or main spar broken or controls smashed or damaged by fire, must not be interfered with unless:

(a) It is necessary to extricate persons or animals from the wreckage.

(b) Further damage may occur unless it is moved.

(c) The aircraft is a danger or an obstruction to the public.

(d) The authority of the Accident Investigation Branch has been given.

Passengers' baggage may be removed from a crashed aircraft provided a police officer is present; in the case of an aircraft coming from abroad, Customs and Excise must also be informed.

Airmiss Reporting Procedure

Whenever a pilot considers that his aircraft may have been endangered by the proximity of another aircraft so that there was a real risk of collision, the incident must be reported initially to the Air Traffic unit with which he is in communication at the time or immediately on landing. The report must be confirmed in writing (Form CA 1094).

The Flight Plan

A pilot may file a flight plan (CA Form 48) at any time but when it is necessary to co-ordinate the flight with other traffic a flight plan is mandatory. More specifically a flight plan must be completed under the following circumstances:

1. For all flights into controlled airspace which is designated Permanent IFR (Rule 21).
2. For all flights in controlled airspace in IMC or at night except when a Special VFR clearance has been obtained and for certain Special Rules Areas/Zones.
3. Flights to and from the UK which cross the UK FIR boundaries.
4. If the pilot wishes to make use of the air traffic advisory service mentioned on page 12.

When it is intended to fly more than ten miles from the coast, over mountainous country, or sparsely populated areas, pilots are strongly advised to file a flight plan so that in the event of any circumstance preventing the arrival of the aircraft at its destination, search and rescue action can be set in motion without delay.

Even when a flight plan is unnecessary pilots must inform air traffic control at the airfield of departure before commencement of the flight. This is known as **Booking Out**. No details of the flight are transmitted to other ATS units.

When a flight plan has been filed and the pilot lands at an airfield which is not that specified in the flight plan, e.g. an **Alternate**, the pilot must advise ATC on arrival.

Carriage of certificates and licences

A pilot flying an aircraft within the UK is required to carry the following documents:

Pilot's licence
Certificate of Airworthiness
Certificate of Registration
R/T licence covering the radio installation in the aircraft.

If, however, the flight is to commence and terminate at the airfield of departure and does not involve the crossing of a foreign territory then these documents may be kept at the home airfield.

Customs and Excise

Flights abroad

A pilot flying abroad is required to have the following documents:

Pilot's licence
Certificate of Airworthiness
Certificate of Registration
Licence covering the radio installation in the aircraft
Passport
Customs form C42
General declaration

It is also advisable to carry an internationally recognized credit card and/or a fuel carnet (fuel credit card issued by the oil companies) and documents relating to public health, e.g. vaccination certificates.

Availability of Customs Facilities

Customs facilities at airports are listed in the following categories related to their hours of attendance

Permanent	Attendance at all times.
Designated	Attendance during airport hours of operation.
Business User	Attendance by prior notice.
Concessional	Attendance by prior arrangement.

Some relevant publications

The Air Navigation Order and Aeronautical Information
Statutory information relating to aviation is published in the Air Navigation Order (ANO) HMSO.

CAP 32 Air Information Publication (*Air Pilot*) provides information on aeronautical facilities, regulations, ground organization and interpreting the requirements of the ANO. The AIP is currently published in three volumes grouped in sections:

GEN   General AGA (aerodromes)
COM   Communications and MET (Meteorology)
RAC   Air Traffic Rules and Services
FAL   Facilitation SAR (Search and Rescue)
MAP   Aeronautical Charts

Notams
The AIP is supplemented by these notices which may be of a temporary nature, (e.g. navigation warnings) or permanent, in which case they will be incorporated in the AIP in due course. Notams are published as required and numbered consecutively as they are issued throughout the year.

Aeronautical Information Circulars
Issued weekly by the UK Civil Aviation Authority notifying administration and operational occurrences. Obtainable from The Aeronautical Information Service, Pinner, Middx.

AUSTRALIA
Air Navigation Regulations (Parts V, IX, X, XI, XII, XIII, XVI)
Air Navigation Orders (Part 40)
Visual Flight Guide
Flight Radio Operator's Manual

CANADA
Air Regulations and Aeronautics Act
Air Navigation Orders
Personnel Licensing Handbook, Vol 1
Flight Information Manual

SOUTH AFRICA
Air Navigation Regulations, 1963

UNITED STATES OF AMERICA
Federal Aviation Regulations (Parts 1, 61, 71 and 91)
National Transportation Safety Board Procedural Regulations (Part 430)
Airman's Information Manual (5 Parts)

Aviation Law Syllabus requirements for the UK Private Pilot's Licence

From the 1st January 1987 the syllabus for Aviation Law – Flight Rules and Procedures is being extended requiring a broader knowledge of the Air Navigation Order 1985 and rules of the Air and Air Traffic Control Regulations 1985, in particular:

Rules of the Air and Air Traffic Control Regulations 1985
Rule
37   Use of Radio Navigation Aids
38   Special Rules for Cross-Channel Air Traffic
39   Special Rules for Air Traffic in the Upper Flight Information Regions
40   Special Rules for Air Traffic in the Scottish Terminal Control Area

Air Navigation Order 1985

Article	Sub-Section(s)	Article	Sub-Section(s)
3		37	
4	6 10 11 12 13	38	
7		39	
8	2	40	
9	1		
10	1 2	41	
11	1 4	45	
13	1 2	51	
14	1	63	1 2 3 4
15		64	2 3 4 5
16		69	
18	1	70	
19	1 6 7	71	1 2
23		82	3 4
33		90	
35	1 2 3 4 5	96	

Schedules
1 (Part A) 2 3 5 6

Personal log books

All pilots are required to keep a personal log book in which the following particulars are recorded:

The name and address of the holder of a log book.

Particulars of holder's licence (if any).

The name and address of his employer (if any).

Particulars of all flights made as a member of the flight crew of aircraft, including:

(a) date, duration and places of arrival and departure of each flight;
(b) type and registration marks of aircraft;
(c) capacity in which holder acted in flight;
(d) particulars of any special conditions under which the flight was conducted, including night flying and instrument flying;
(e) particulars of any test or examination undertaken while in flight;
(f) details of simulator training including details of type, practice and times.

Questions

1 **An applicant for a Private Pilot's licence must have completed a course which includes at least:**
   (a) 20 hours dual instruction and 10 hours solo,
   (b) 23 hours dual instruction and 20 hours solo,
   (c) 20 hours dual instruction and 18 hours solo.

2 **A Private Pilot's licence is issued for:**
   (a) One year,
   (b) Three years,
   (c) Permanently.

3 **An aircraft in group 'B' is:**
   (a) A single-engine aircraft above 5700 kg/MTWA
   (b) A single-engine aircraft below 5700 kg/MTWA
   (c) A multi-engine aircraft below 5700 kg/MTWA

4 **The holder of a Private Pilot's Licence may:**
   (a) Fly passengers for hire or reward, only within the UK.
   (b) Fly passengers anywhere in the world but not for hire or reward.
   (c) Fly passengers for hire or reward outside the UK.

5 In order to maintain a group on a Pilot's Licence without having to take a flight test the holder must have flown an appropriate aircraft within the previous 13 months for not less than:
   (a) 5 hr,
   (b) 10 hr,
   (c) 10 hr of which 5 hr may be in a simulator.

6 To obtain a night rating a pilot with the required day flying experience must complete at least:
   (a) 5 hr instrument instruction and 5 hr night flying under supervision,
   (b) 5 hr instrument instruction an 10 hr night flying under supervision.
   (c) 10 hr instrument instruction and 5 hr night flying under supervision.

7 When two aircraft are approaching head on:
   (a) Each aircraft shall alter heading to the left,
   (b) The smaller aircraft shall alter heading to the right,
   (c) Each aircraft shall alter heading to the right.

8 When two aircraft are flying on converging headings:
   (a) The one which has the other on its port side shall give way,
   (b) Both shall turn onto diverging headings,
   (c) The one which has the other on its starboard side shall give away.

9 Unless ATC has allocated priority, when two aircraft are approaching to land at the same time:
   (a) The one which has the greater height must give way to the lower aircraft,
   (b) The larger aircraft has right of way,
   (c) The one nearest the runway threshold has right of way to land.

10 Above 3000 ft outside controlled airspace a pilot must fly within the following weather conditions to remain VMC:
   (a) 1 mile horizontally and 1000 ft vertically from cloud and 3 n.m. visibility,
   (b) 1 mile horizontally and 1000 ft vertically from cloud and 5 n.m. visibility,
   (c) 1 mile horizontally and 500 ft vertically from cloud and 5 n.m. visibility.

11 **In the UK flight at night under VFR:**
   (a) Is not permitted,
   (b) May be permitted,
   (c) Is permitted only when the pilot has an instrument rating.

12 **An area where firing and bombing practice is permanently active by day and night is indicated on a map by:**
   (a) A pecked or dotted red outline,
   (b) A solid blue outline,
   (c) A solid red outline.

13 **Using the quadrantal rule a pilot making good a magnetic track of 180° would fly:**
   (a) Odd thousands of feet + 500 ft,
   (b) Even thousands of feet + 500 ft,
   (c) Even thousands of feet.

14 **When flying under VFR, flight separation is the responsibility of:**
   (a) ATC in the Flight Information Region,
   (b) The pilot,
   (c) ATC Service on the frequency being worked.

15 **Other than during a landing or take-off, normally an aircraft may not fly close to persons or property. This limit is:**
   (a) 500 ft,
   (b) 1000 ft,
   (c) 2000 ft.

16 **For flights abroad certain aircraft documents are required. These must include:**
   (a) C of A, General Declaration, Certificate of Maintenance,
   (b) General Declaration, Fuel Carnet, C of A,
   (c) C of A, Radio Installation Licence, General Declaration.

17 **Two red balls on the signals mast coupled with a double cross in the signals area denotes:**
   (a) Parachute dropping is in progress,
   (b) Instrument Flight Rules in force,
   (c) Glider flying is in progress.

18 **An intermittent white beam directed to an aircraft on the ground indicates:**
  (a) Expedite take-off,
  (b) Return to the parking area,
  (c) Flight plan has been cancelled.

19 **A pilot may file a flight plan at any time but it is mandatory:**
  (a) If it is intended to fly over the sea,
  (b) If radio is not installed in the aircraft,
  (c) For flight at night.

20 **The UK is divided into the following Flight Information Regions:**
  (a) London and Scottish,
  (b) London, Midland and Scottish,
  (c) Southern, Preston and Northern.

21 **The Lower Airspace Advisory Service is normally available to:**
  (a) all aircraft
  (b) all aircraft in controlled airspace up to FL75,
  (c) all aircraft in uncontrolled airspace up to FL95.

22 **At night a military airfield may be identified by a beacon flashing in morse:**
  (a) The airfield letters in green,
  (b) The airfield letters in red,
  (c) The airfield letters in white and red.

23 **When using a prominent line feature for navigational purposes the pilot should:**
  (a) Fly to the right of the feature,
  (b) Fly overhead the feature,
  (c) Fly to the left of the feature.

24 **Flying instruction for the purpose of gaining a licence or rating may be given by:**
  (a) Any pilot with a professional licence,
  (b) A qualified flying instructor,
  (c) Any licenced pilot so long as no payment is made.

25 Shortly before a night landing the radio fails and visual signals have to be used. The pilot sees a steady red beam directed at him from the ground. This means:
  (a) Land at another airfield,
  (b) Return to your point of take-off,
  (c) Landing temporarily suspended; wait.

26 With certain exceptions Transition Altitude in the UK is:
  (a) 4000 ft amsl,
  (b) 3500 ft amsl,
  (c) 3000 ft amsl.

27 While flying at night you see the green navigation light of another aircraft flying on a similar heading at the same height as your aircraft. It appears to be coming closer. What action would you take?
  (a) Hold the present height and heading but be ready to take avoiding action if this is required.
  (b) Alter heading to starboard,
  (c) Climb.

28 The UK Department of Transport Accident Investigation Department must be notified when an accident involves:
  (a) A forced landing due to engine failure,
  (b) Damage in a hangar during maintenance,
  (c) Damage due to a technical defect in the aircraft.

29 While taxying back to the parking area you see a vehicle towing an aircraft moving towards your intended path. What action should you take?
  (a) Proceed on your present direction since you have right of way,
  (b) Turn right,
  (c) Take such avoiding action as is appropriate. The towing vehicle has right of way.

30 A detailed explanation of UK aviation law is given in:
  (a) The Air Navigation Order and Rules of the Air and ATC Regulations,
  (b) The Air Information Publication,
  (c) Aeronautical Information Circulars.

Chapter 2
Navigation

Form of the Earth, Meridians of Longitude, Parallels of Latitude

For convenience maps are produced on flat sheets of paper. However, the earth's surface is spherical and it is impossible to make a flat sheet conform to the double curvature of a sphere. It is equally impossible to flatten a spherical surface without distortion. By depicting relatively small areas of the earth on a map these errors are minimized, but the navigator, requiring a high degree of accuracy, cannot be satisfied with these steps alone.

Various methods of projecting the earth's surface on to a flat sheet have been evolved over a great number of years. These **Projections** as they are called represent a considerable study in themselves

Broadly speaking maps fall into two categories:

1. Topographical: used for map reading.
2. Navigational: designed for a method of plotting known as **Dead Reckoning (DR) Navigation**. This was the method adopted by the specialist navigators and it is now more or less extinct.

Whereas maps refer to land areas, charts depict water although the two terms have become interchangeable.

So far as the Pilot-Navigator is concerned, topographical maps are the correct tool for the task since navigational sheets offer little detail for map reading and their method of projection is designed primarily to meet the requirements of DR navigation.

A good topographical map should offer the following characteristics:

(*a*) Area should be accurate.
(*b*) Shape should be correct.
(*c*) A line drawn on the map should represent the shortest distance between two points on the earth's surface.

Unfortunately it is not possible to attain all of these requirements to perfection on one sheet and some ingenius projections have been developed for topographical maps which represent a compromise. Whereas one projection may be ideal for a north–south country such as the British Isles, another is more suitable for an east–west country like the United States of America. Modern maps are masterpieces of design and printing, providing the pilot with a high degree of all-round accuracy.

For reference, the earth is covered by a network made up of **Meridians of Longitude** and **Parallels of Latitude** (Fig. 9).

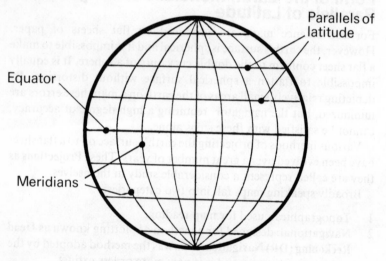

Fig. 9   Meridians of Longitude and Parallels of Latitude.

Meridians of Longitude

The famous **Prime Meridian** running through Greenwich is designated zero, all others being referred to as east or west of Greenwich, e.g. 10° W; 45° E; 179° W, etc. On aviation maps meridians are usually shown at 10′ (**Minute**) intervals, there being 60′ to a degree, so that a more accurate meridian can be quoted, e.g. 10° 30′ W; 45° 15′ E; 179° 59′ W, etc. When higher levels of accuracy are required decimals of a minute are used. An older method, now rare in aviation is to quote an accurate position in degrees, minutes and **Seconds** (″).

180° is where east meets west, this meridian being opposite to the Greenwich meridian on the other side of the earth. It, too, is of special interest being the **International Date Line** where, for example, it can be Monday on one side and Tuesday on the other. Fortunately domestic complications are avoided since meridian 180° passes over the Pacific Ocean until it reaches the thinly populated north-east tip of Russia.

Parallels of Latitude

The Equator is the most commonly known parallel, all others being numbered north or south of it, e.g. 57° N; 22° S; 80° N. Like meridians these are shown at 10' intervals on aviation maps. 90° N and 90° S are the north and south poles respectively.

Using the network created by meridians and parallels any point on the earth can be quoted by giving its **Latitude** and **Longitude** (Lat. and Long.), for example, Kidlington Airfield is 51° 50' N, 01° 19' W.

The complete network made up of parallels and meridians is called a **Graticule**. In addition to providing a means of reference, the graticule is used to position the protractor when measuring direction in relation to true north, all meridians running north–south and parallels east–west.

Some maps are overprinted with an alternative graticule called the **British Modified Grid**. This is used by the Army and is unsuitable for the pilot-navigator since the Grid bears little relation to geographical N, S, E, or W. Care should be taken when purchasing maps to see that they are overprinted with latitude and longitude.

Great Circles and Rhumb Lines

Assuming it is required to fly from Le Touquet airport to Ostend airport, a line is drawn on the map joining these two places. This is called the **Required Track**, referred to in abbreviated form as **Tr. Req.** The bearing of the Tr. Req. is measured with a protractor positioned on any parallel of latitude or meridian which crosses the track, but certain precautions must be taken. It will be remembered that the meridians converge towards the poles. This can be confirmed by measuring the distance between meridians at the foot of a topographical map and comparing the measurement with the distance between the same two at the top of the sheet. Fig. 10 shows that because of convergence towards the poles tracks running in an easterly–westerly direction will cross each meridian at a slightly different angle. Obviously a north–south track is not similarly affected. The protractor should therefore be positioned on the meridian or parallel which crosses nearest the centre of track to

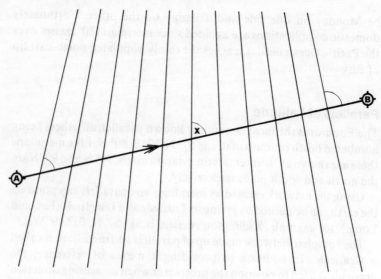

Fig. 10 Measuring a westerly/easterly track. Compare the angle at the meridian near airfield A with that near airfield B and note each meridian crosses track at a different angle. To obtain an average reading it should be measured at point x.

obtain the average angle. On topographical maps such a track represents the shortest distance between two points on the earth's surface and it is called a **Great Circle**.

Long-distance navigators prefer tracks which cross each meridian at the same angle and these are called **Rhumb Lines**. Special charts, distorted towards the poles, are produced for the purpose. They are called Mercator's Projections and all lines drawn on them are rhumb lines. These are essentially navigational projections, unsuitable for the pilot-navigator and this passing reference will suffice.

Aeronautical Maps and Symbols

Although many specialist maps are available, for the purpose of pilot navigation based on map reading three types are most widely used. In the main these differ in scale, a subject to be explained in the next section.

For a map to be of value it must depict features of importance to the pilot. These are often not the same as those likely to interest a motorist because, from the air, features that might be difficult to recognize could dominate the vision while driving a car. The relative importance of ground features while map reading from the air will be discussed later but, at this stage, it should be clear that railways, lakes, rivers, woods, etc., all of them shown in a school atlas, would not provide sufficient information for a pilot. Neither would they be depicted in the correct degree of emphasis for recognition from the air.

For pilot navigation the map must show airfields/airports along with their elevation above sea level, spot heights of tall structure (TV masts, etc.), high ground, danger/restricted areas which have to be avoided, controlled airspace which must likewise not be violated and boundaries of the Flight Information Regions (FIR) described in Chapter 1. Also, the location of radio facilities intended for air navigation is usually included.

The various features mentioned are depicted on maps by symbols which are known as **Conventional Signs**. These are illustrated in Fig. 11. There was a time when symbols for the same ground feature differed from one series of maps to the next but in recent years there has been a move towards standardization with the exception that maps of larger scale show the runway layout of airfields in place of the circular symbol printed on those of smaller scale.

Scale, units of measurement and conversion

A map is, in fact, a scale drawing of a particular area of the earth. Those found in a school atlas are produced to show a large area on a relatively small sheet and a scale of 100 miles or more to the inch is common. At the other extreme are the maps used by surveyors and sometimes hikers, which indicate a mile for every $2\frac{1}{2}$ in. on the sheet. Such maps are very detailed, showing quite small buildings and every minor bend in a country lane. Aviation requirements fall between these two examples and those in common use are:

(*a*) The 1:250,000 series
(*b*) The 1:500,000 series
(*c*) The 1:1,000,000 series.

They are usually referred to as the 'quarter-million', 'half-million', or 'million series' respectively. As the terms imply, one unit on, for example, a half-million map denotes 500,000 units on the earth. For practical purposes this amounts to eight statute miles to the inch but accurate scales are printed on the map depicting statute miles, nautical miles and kilometres.

AERODROMES

Field limits with runway pattern

Howes-Name   168 -Elevation   (M)-Military, (C)-Civil (J)-Joint

Howes
168
(M)

Field limits with runway pattern unknown, or not portrayable

Minor aerodromes with runway pattern unknown, or not portrayable

Disused or abandoned aerodromes

Heliports (M)-Military, (C)-Civil, (J)-Joint

Glider Launching Site (See UK AIP RAC 5-1-3)
a. Primary activity at locations
b. Additional activity at locations

Parascending Site (See UK AIP RAC 5-1-7)
a. Primary activity at locations
b. Additional activity at locations

Hang-Gliding Site (See UK AIP RAC 5-1-7)

Free-fall Parachuting Site, DZ circle 1¼ NM radius

Site of Intensive Microlight Flying
(Intensive Microlight activity also takes place at certain Licenced and Unlicenced)

Special Rules Zone (SRZ) SFC-FL65

Special Rules Area (SRA) .... 1500'ALT-FL65

Control Zone (CTR) .... SFC-2500'ALT

Control Area (TMA and CTA)
Outer Boundary .... 3000'ALT-FL245
Inner Boundary .... 1500'ALT-FL245

Control Area (Airway) .... R 3 2500'ALT-FL245

Exceptionally High Obstruction (Lighted) Single, Multiple 1000ft. or more AGL 1950 (1720)   1536 (1300)

Obstruction (Unlighted) 530 (323)

Group Obstruction (Lighted) 560 (425)

Numerals in italics indicate height of top of obstruction above Mean Sea Level. Numerals in brackets indicate height of top of obstruction above Local Ground Level.

Cables joining Obstructions, height AGL 310

SYMBOLS ARE NOT SHOWN ON THIS CHART FOR LAND-SITED OBSTRUCTIONS LESS THAN 300FT ABOVE LOCAL GROUND LEVEL. PERMANENT OFF-SHORE OIL AND GAS INSTALLATIONS ARE SHOWN REGARDLESS OF HEIGHT CATEGORY.

Aeronautical Light ...... ☆ F l R

Power transmission line
(over 200ft AGL)
Powerline information is not necessarily complete.

Visual Reporting Point (VRP) ...... ASHFORD

Special Access Lane Entry/Exit ...... E
(indicates centre of lane, see UK AIP for conditions of use)

Altimeter Setting Region ...... E CHATHAM PORTLAND

Bird Sanctuary ......
Areas are shown with name/effective altitude (in thousands of feet AMSL).

High Intensity Radio Transmission Area
Areas are shown with name/effective altitude (in thousands of feet AMSL).

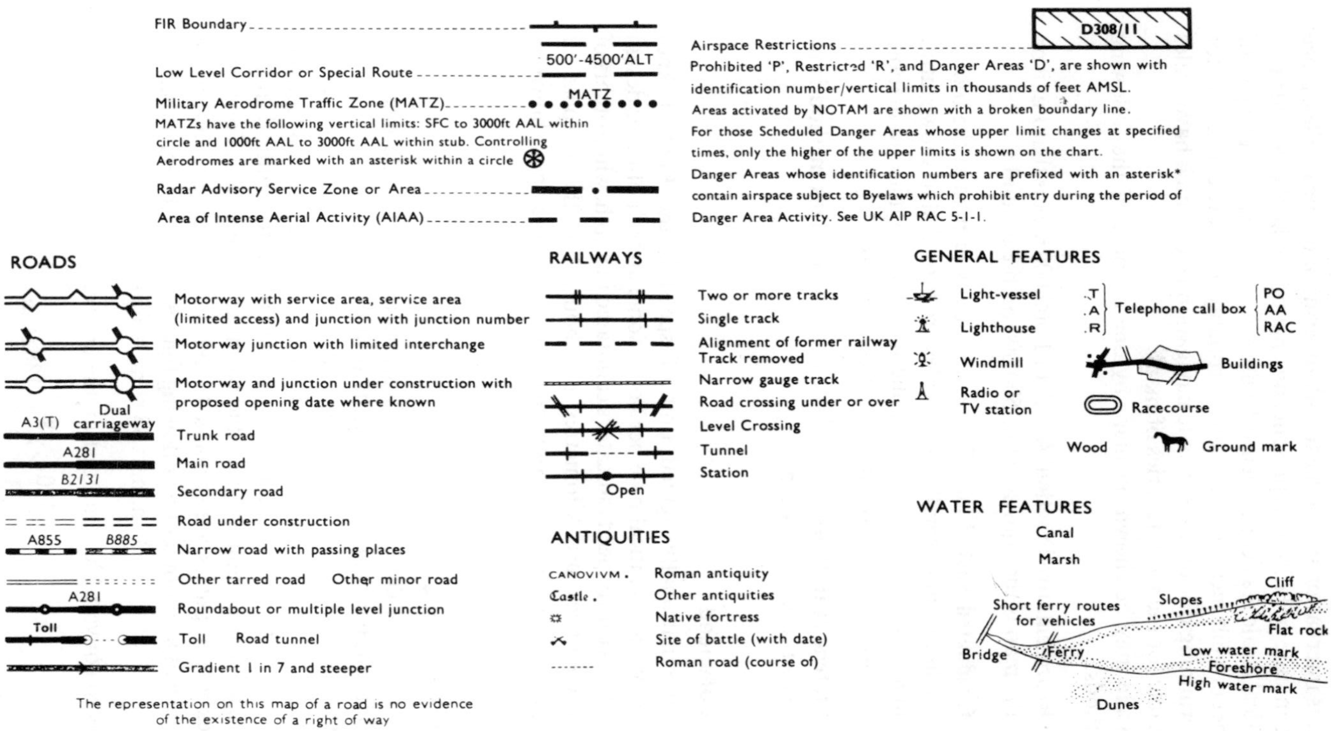

D308/11

FIR Boundary

Low Level Corridor or Special Route — 500'-4500'ALT

Military Aerodrome Traffic Zone (MATZ). — MATZ
MATZs have the following vertical limits: SFC to 3000ft AAL within circle and 1000ft AAL to 3000ft AAL within stub. Controlling Aerodromes are marked with an asterisk within a circle

Radar Advisory Service Zone or Area

Area of Intense Aerial Activity (AIAA)

Airspace Restrictions
Prohibited 'P', Restricted 'R', and Danger Areas 'D', are shown with identification number/vertical limits in thousands of feet AMSL. Areas activated by NOTAM are shown with a broken boundary line. For those Scheduled Danger Areas whose upper limit changes at specified times, only the higher of the upper limits is shown on the chart. Danger Areas whose identification numbers are prefixed with an asterisk* contain airspace subject to Byelaws which prohibit entry during the period of Danger Area Activity. See UK AIP RAC 5-1-1.

GENERAL FEATURES

Light-vessel   Telephone call box {PO AA RAC}
Lighthouse
Windmill   Buildings
Radio or TV station
Wood   Racecourse   Ground mark

RAILWAYS

Two or more tracks
Single track
Alignment of former railway Track removed
Narrow gauge track
Road crossing under or over
Level Crossing
Tunnel
Station   Open

ANTIQUITIES

CANOVIVM · Roman antiquity
Castle ·   Other antiquities
Native fortress
Site of battle (with date)
Roman road (course of)

WATER FEATURES

Canal
Marsh
Short ferry routes for vehicles
Bridge Ferry
Cliff Flat rock
Slopes
Low water mark Foreshore
Dunes High water mark

ROADS

Motorway with service area, service area (limited access) and junction with junction number
Motorway junction with limited interchange
Motorway and junction under construction with proposed opening date where known
A3(T) Dual carriageway
A281 Trunk road
B2131 Main road
Secondary road
A855 B885
Road under construction
Narrow road with passing places
A281 Other tarred road Other minor road
Roundabout or multiple level junction
Toll Toll Road tunnel
Gradient 1 in 7 and steeper

The representation on this map of a road is no evidence of the existence of a right of way

Fig. 11 Map symbols. Most are common to both the 1:250,000 and 1:500,000 series.

39

It is international practice to talk in terms of nautical miles. Since one minute of latitude represents one nautical mile the latitude (i.e. vertical) scale may conveniently be used in conjunction with a pair of dividers while measuring distance. Probably still more convenient for determining distance are plastic navigation rules which have scales suitable for the various series of maps.

Scale may be shown on maps by one or more of the following methods:

1. **Representative Fraction**. An r.f. of 1:500,000 means that 1 unit on the map is the equivalent of 500,000 on the ground.

2. **Graduated Scale Lines**. These are usually printed at the foot of the map and normally three scales are provided to show statute miles, nautical miles and kilometres. Nautical mile graduations are printed on meridians at degree intervals.

3. **Statement in Words**. e.g. '¼ inch to 1 mile'. This method is no longer used in aviation.

Units of speed

The internationally agreed unit of speed is the **Knot**, 1 knot representing 1 nautical mile per hour. Some aircraft of older design are fitted with airspeed indicators calibrated in m.p.h. while others of some countries use kilometres per hour. For the purpose of navigation it is obviously convenient to work in the units shown on the airspeed indicator.

Unit conversions

A statute mile contains 5280 ft, whereas a nautical mile measures 6080 ft. The nautical mile represents an attempt to relate a measuring unit to the dimensions of the earth and it is described as 'the average length of one minute of arc'. It may be necessary to convert kilometres to statute miles or statute miles to nautical miles. Navigational computers are able to make these conversions very quickly (computers are explained later in the chapter), but in the absence of a computer the following figures are worth remembering:

$$33 \text{ n.m.} = 38 \text{ st.m.} = 61 \text{ km.}$$

Thus to convert 120 n.m. to st.m. the answer is

$$\frac{120}{33} \times \frac{38}{1} = 138 \text{ st.m.}$$

or 250 st.m. converted to n.m. is

$$\frac{250}{38} \times \frac{33}{1} = 217 \text{ n.m.}$$

Heights, Units of Measurement and Conversion

The international unit of height measurement for air navigation is feet and most altimeters are so calibrated. However, some European countries continue to publish maps giving heights in metres. In either case the heading of the map will state the units of height being used.

Unit conversions

A height conversion table, comprising metres and feet scales printed side-by-side, is provided in the side or bottom margin of the map. Alternatively the navigation computer will rapidly make conversions between units.

For very approximate purposes, heights in feet can be found by multiplying the figure shown in metres by 3. Conversely, a height in feet may be converted to metres when divided by 3. An accurate conversion, one quickly made with the aid of an electronic calculator, is as follows:

To convert feet into metres   multiply by 0.3048

To convert metres into feet   multiply by 3.2808

Methods of Depicting Relief

Although high ground is not always of value as a map-reading feature, a prominent hill or mountain can present an excellent pinpoint. Furthermore changes in depth when flying over water can sometimes be seen quite clearly. In addition, high ground presents a hazard in conditions of poor visibility, low cloud and darkness. For all of these reasons, relief is of interest to the pilot-navigator and the various means of showing it on a map are listed below:

1.   Spot heights (and depths)
2.   Layer tints
3.   Contours
4.   Form lines
5.   Hill shading
6.   Hachures.

1. **Spot Heights** indicate the higher points, usually of a prominent nature, which have been measured above **Mean Sea Level (MSL)**.

Man-made obstructions, such as television/radio masts, are shown with their heights above sea-level and, in brackets, heights above ground.

2. **Layer Tints** are used in conjunction with contour lines on topographical maps. The palest shades refer to the lower regions, white being reserved for ground at sea-level. The tints become darker as higher ground is indicated and a chart showing heights against each shade is printed on the margin of the map.

3. **Contours** are lines joining all places of equal height. With practice they give a good indication of the shape of a hill or mountain. Widely spread, they denote a gentle rise or fall, whereas a steep-sided mountain will be depicted as a series of closely spaced contour lines. At convenient intervals the lines are broken and the height for that particular contour is printed in the resultant space.

4. **Form Lines** are similar to contours except that they are an approximation in the form of concentric rings rather than the accurate picture of rising ground presented by contour lines.

5. **Hill Shading** is not in general use on aviation maps these days, partly because it tends to obliterate other details. It represents an attempt to endow the map with a three-dimensional effect, by shading those portions of the high ground which would lie in the shadows when the sun is in a certain position.

6. **Hachures** are not to be found on topographical maps. In this method the approximate outline of a range of hills is shown as a row of short, fence-like lines. Hachures may be seen on navigational plotting sheets and their purpose is to indicate the presence of high ground rather than attempt to describe it.

The pilot-navigator need only be concerned with those methods which are common to topographical maps, namely spot heights, layer tints and contour lines. A combination of all three is usual.

Headings: Relationship between True, Magnetic and Compass North

Although the aircraft is steered with reference to the direction indicator this gyroscopic-operated instrument as fitted to training and light touring aeroplanes must first be aligned with the magnetic compass before its readings have any navigational meaning. However, there is a difference between geographic north or south as seen on a map and the north or south indicated by the magnetic compass.

Variation

In the paragraph dealing with latitude and longitude it was explained that the north and south poles are the main reference points.

Meridians extend towards the poles, and any direction measured on a map will be a **true** direction. However, a compass needle influenced solely by the earth's magnetic field does not point towards **True North**, but instead takes up a direction towards **Magnetic North** – a position in Baffinland, north-eastern Canada many miles from the north geographical pole. Furthermore, this magnetic pole is not fixed but is moving very gradually round the true pole. Thus the direction indicated by a compass needle will alter each year in relation to true north, although the change is very small. In Great Britain the annual change is about 7′ east. The difference, between the direction of true north and magnetic north is known as **Variation**.

The amount of angular difference alters from place to place and may be east or west of true north. For example, the variation in Burma and Western Australia is nil, whereas at Vancouver Island it is 25°E and on the other side of Canada, in Newfoundland, it is over 30°W.

To sum up: variation is the angular difference between the true meridian and the magnetic meridian. The size of this angle depends on:

(*a*) Position on the earth's surface;

(*b*) The year, because the magnetic pole is not static. (The date of the variation and the annual change is stated on all maps.)

The method of showing variation on maps may be in the form of:

(*a*) **Isogonals** – chain lines joining places that have equal variation

(*b*) Diagram

(*c*) Statement in words

(*d*) Compass rose.

Deviation

When a compass is installed in an aircraft it ceases to be influenced solely by the earth's magnetic field. Local magnetism, from electrical wiring and iron and steel components within the aircraft, has a disturbing effect and causes the compass needle to point towards **Compass North** and not Magnetic North. The angular difference between these two directions is known as **Deviation**. It is expressed as 'plus' or 'minus'.

During a turn the magnetic materials within the aircraft change their position relative to the compass needle, or, in other words, the aircraft moves but the needle remains pointing to compass north. Consequently, deviation will alter with the heading of the aircraft; e.g. on a heading of 090° deviation it may be +3° and on a reciprocal (270°) it may be —3°, while at some point between these headings it would be nil. The aircraft compass is fitted with adjustable compensating magnets positioned in a **Corrector Box** attached to the underside of the compass bowl. By manipulating the magnets with a key it is possible to reduce variation to a minimum, but it cannot be eliminated entirely. For the pilot's information, therefore, a card is tabulated with the amount of deviation remaining after correcting adjustments have been made. This **Deviation Card** is mounted in a frame positioned near the compass.

The Compass Swing

At intervals, and whenever the aircraft has flown near a thunderstorm, a new deviation card is compiled. A **Compass Swing** is enacted on the ground and away from metal-framed buildings. On some airfields there is a suitably marked compass swinging base but more usually a surveyor-type compass is used to measure aircraft heading. The aircraft compass is then adjusted via the corrector box and the small, residual errors are listed on the deviation card.

Allowing for Variation and Deviation

Consider what must be done to convert a true heading – Hdg (T) – into a compass heading – Hdg (C) – or, in other words, how is the allowance made for both variation and deviation?

Having found variation and deviation it only remains for the pilot to determine whether the figures should be added or subtracted to the true heading to find Hdg (C), i.e. the heading to steer on the compass. The simple rule of application for variation may be remembered by the words

'East is least and West is best.'

In other words, to convert true to magnetic, deduct easterly variation and add if westerly. The deviation card provides corrections as a + or — figure which should be applied accordingly to change magnetic to compass heading.

TRUE VIRGINS MAKE DULL OMPANY

TRue VAR   MA9   DEV Compass

Figure 12 illustrates the procedure. Hdg (T) is shown as the angle
between true north and the direction in which the aircraft is heading –
in this case 128°. The second sketch shows 10°W variation added to
Hdg (T) to produce a Hdg (M) of 138°. From the correction card in
the cabin deviation on this heading is +3°, i.e. three degrees west of
magnetic north. Further reference to Fig. 12 will show that the +3°
deviation is added to obtain the correct Hdg (C) —141°.

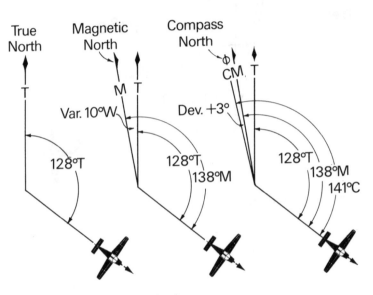

Fig. 12   Variation and Deviation b.

When converting Hdg (C) to Hdg (T) calculations have to be
reversed.
Complete the following examples for practice:

	Hdg(T)	Var.	Hdg (M)	Dev.	Hdg (C)	
(1)	108°	10°W	?	−2°	?	(Fig. 13)
(2)	324°	13°W	?	+1°	?	
(3)	?	5°E	?	+3°	052°	(Fig. 14)
(4)	?	8°W	?	−1°	183°	
(5)	?	16°W	279°	−1°	?	(Fig. 15)
(6)	?	5°E	358°	+3°	?	

Answers:
(1) Hdg (M), 118°; Hdg (C), 116°.
(2) Hdg (M), 337°; Hdg (C), 338°.
(3) Hdg (M), 049°; Hdg (T), 054°.
(4) Hdg (M), 184°; Hdg (T), 176°.
(5) Hdg (T), 263°; Hdg (C), 278°.
(6) Hdg (T), 003°; Hdg (C), 001°.

45

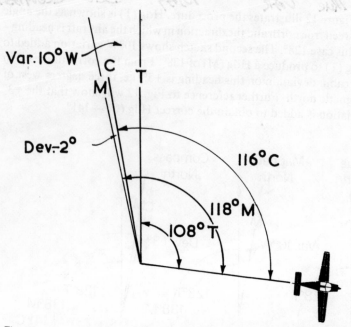

Fig. 13 Finding Hdg(M) and Hdg(C).

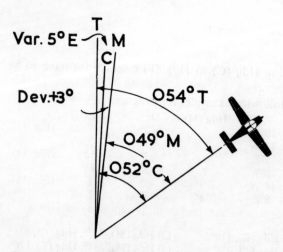

Fig. 14 Finding Hdg(M) and Hdg(T).

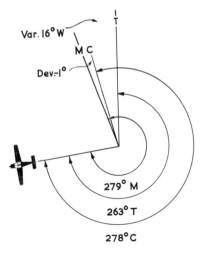

Var. 16° W

M C

Dev.-1°

279° M

263° T

278° C

Fig. 15   Finding Hdg(C) and Hdg(T).

Earth's Magnetic Field, Isogonals and the Aircraft Compass

The earth's magnetic field has for many years been plotted and recorded with some accuracy, thus enabling the map maker to depict local variation in a form useful to the pilot. Most usually this takes the form of isogonals, the lines previously mentioned which join places of equal magnetic variation.

The earth is a magnet of massive proportions and its north and south behave like any other magnet, however small. Although the magnetic compass seeks north or south by the mutual attraction of its own, small magnetic field and that of the earth there are certain side effects or errors which must be understood if the instrument is to be used effectively.

Factors affecting the Compass

Dip

For the compass scale to rotate freely within the instrument it must be suspended on a needlepoint bearing of minimum friction. However, the earth's magnetic field, emanating from within, attracts the compass magnet system *downwards* as well as towards the poles. Figure 16 shows that, at the equator the downward pull exerted by the

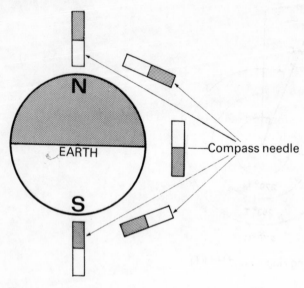

Fig. 16   The effects of "Dip". Shaded areas are Blue poles of the magnet, clear areas are Red. Unlike magnetic poles attract, consequently the earth tends to pull down the compass needle as an aircraft flies towards the north or south pole.

north and south poles is equal and the compass magnet system will remain level; in equatorial regions there is no dip.

As one moves towards either pole the north- (or south as the case may be) seeking end of the compass magnets is progressively pulled down until, at the poles, maximum dip occurs and the magnet system will stand on end.

The effects of dip are, to some extent, minimized by suspending the compass magnet system below its pivot so that it acts like a pendulum but a small amount of dip remains and, under certain conditions of flight, it can distort the readings of the instrument. There is no need for a pilot to have a deep understanding of these errors but a working knowledge is important to allow accurate setting of the direction indicator.

Acceleration and Deceleration

The following explanations relate to flights in the northern hemisphere and the errors described will act in reverse south of the equator.

Imagine the aircraft is flying on an easterly or westerly heading. If, for any reason the aeroplane accelerates there will be a tendency for the magnet system (which is suspended below the pivot point) to be left behind. Dip will attempt to pull down the north-seeking end of the magnet system and cause the compass to read an apparent turn towards north although the heading has remained unchanged.

Now consider a deceleration. In this case the magnet system will tend to move forward of the pivot and the earth's magnetic field will attempt to pull down the compass magnets and cause an apparent turn towards south.

Acceleration and **Deceleration Errors** may convincingly be demonstrated in an aircraft by holding a constant heading on a distant ground feature, then increasing and decreasing speed while noting the effect this has on the compass.

Lateral Level

When the aircraft is flying on a northerly or southerly heading and a wing is allowed to go down, sideslip will cause the magnet system to be displaced, allowing dip to pull down the north-seeking end of the magnet system towards the lower wing.

Precautions while setting the direction indicator

The two compass errors just described, one affecting the instrument on E–W headings, the other on N–S, must be borne in mind while aligning the direction indicator. Otherwise the DI will give misleading readings. When setting the DI **the airspeed must be steady**, and **the wings must be level with the aircraft in balance**.

Turning

Compass errors affect the behaviour of the magnetic compass during turns. Although acceleration/deceleration errors are at their maximum while flying due west or east and lateral level errors are most pronounced when heading north or south, on intermediate headings a proportion of both errors will occur.

What could be a complicated matter is avoided by making turns onto a required heading using the gyroscopic-operated DI. However, if for any reason it is necessary to make turns with reference to the magnetic compass these rules apply in the northern hemisphere:

1. Limit turns to Rate 1. Beyond that rate the compass becomes erratic and unreadable.

2. While turning onto east or west from a northerly or southerly heading roll out 10° before reaching the required compass heading. This will allow time for the turn to stop.

3. When turning onto north from an easterly or westerly heading roll out of the turn 20–25° before reaching the required compass heading. As the wings roll level the compass will swing through the last 20–25°.

4. When turning onto south from an easterly or westerly heading roll out of the turn 20–25° after reaching the required compass heading. As the wings roll level the compass will swing back through 20–25°.

In the southern hemisphere items 3 and 4 are reversed, i.e. overshoot on north and undershoot on south.

Proximity of metal objects

Deviation was explained on page 43. However, even when there has been a compass swing and a deviation card has been compiled the compass will be influenced by metal objects brought into the aircraft. So beware! A bag of metal golf clubs stowed in the cabin can easily put an aircraft many miles off track.

The compass may also be affected by electronic pocket calculators or small portable radio receivers although there is usually no disturbance when they are switched off.

Triangle of Velocities

Navigation would be simple in a world free of winds. Having measured the Track, the pilot could set this heading on the compass after correcting it for variation and deviation (page 44) and provided the aircraft was steered accurately all would be well. However, wind is a feature of this world and it has a profound effect on aerial navigation. Wind is movement of a mass of air in relation to the ground. The aeroplane being supported in an air mass is susceptible to its movements like a goldfish carried in its tank from one room to another. The water is moved and the fish goes with it.

Put into more specific terms, an aeroplane cruising at 100 kt TAS from Detroit Metro to a point 100 n.m. due east would in still air arrive over that point after 60 min. The same flight against a 20 kt wind would reduce the speed over the ground to 80 kt although the airspeed would remain unaltered. The flight would now have an elapsed time of 75 min, i.e. 100 n.m. at 80 kt.

Now consider the flight with a 20 kt tailwind. Speed over the ground would be 120 kt and a flight of 100 n.m. distance would take only 50 min. Figure 17 illustrates the effect of a 20 kt headwind and tailwind and it will be apparent that in these cases, provided the track of 090° is corrected for variation and deviation, then accurately steered as a heading, the aeroplane will arrive over point **B**.

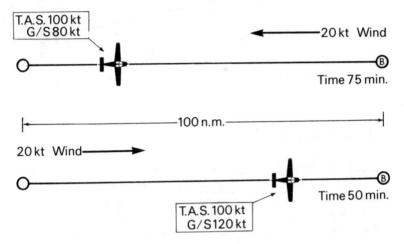

Fig. 17   Effect of a 20 kt headwind and a 20 kt tailwind on ground speed and time of flight.

The situation becomes a little more complex when the wind is at an angle to the track. Taking the same flight as an example with a 20 kt wind from the north, speed over the ground would be little affected, but if 090° is steered at the end of 60 min flying the aeroplane would be 20 n.m. south of track. To arrive over B it is necessary to aim at an imaginary position 20 n.m. north of it thus compensating for the drift caused by the wind (Fig. 18). From this explanation the importance of the wind can be seen, together with the necessity of allowing for its effect in order to make good the required track. Wind strength as well as its direction will affect the aeroplane and the combined factors of wind speed and direction are known as a **Wind Velocity**. Prior to a cross-country flight this can be obtained from the meteorological office or airfield control. It is expressed as a direction and speed, e.g. 045°/25 kt. The direction refers to origin of the wind and not its line

51

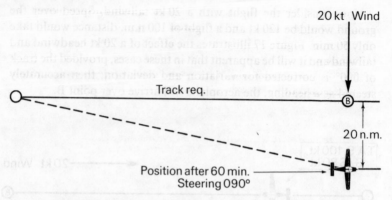

Fig. 18 Wind effect after 60 minutes flying with a 20 kt crosswind blowing at 90 to track.

of travel. The example means that the wind is blowing *from* 045° at 25 kt.

Because the wind strength and direction tend to alter with height, a weather report will include wind velocities at different levels and that quoted nearest to the desired cruising altitude should be used.

With a wind velocity in his possession, the TAS of the aeroplane decided and the Tr. Req. drawn and measured for length and bearing, the pilot is now able to calculate the missing links which are:

1. The true heading (when corrected for variation and deviation; this is steered to compensate for the wind);
2. The ground speed;
3. The elapsed time for the flight.

The Airtour Computer is one of several navigational computers capable of solving such a problem rapidly. There are also some excellent electronic navigation computers available, including the RADNAV 720 which can memorize more than 100 radio facilities and compute tracks, headings, etc. However before using these tools of navigation it is imperative that the student pilot should understand the theory of the **Triangle of Velocities**.

Taking the simple case quoted of a Tr. Req. 090°, TAS 100 kt and a wind velocity (W/V) of 360°/20 kt, it is possible to draw the various speeds and directions (velocities) in relation to one another, using a protractor and a pair of dividers or compasses. Squared paper is preferable but plain will suffice and the following sequence should be followed:

1. Draw the Tr. Req. in the direction 090° but pay no attention to its length at this stage. The conventional sign for track is ➤➤ and it should be marked accordingly to distinguish it from other lines to be drawn later.

2. At the destination end of the track, draw a line to represent the wind direction. Remember this is quoted *from* its origin and in consequence will always blow towards the track. The wind direction should be marked ➤➤➤ which is the conventional sign for a W/V. The wind direction and track have now been drawn in relation to one another.

3. Next insert the wind speed. Any scale can be used for this purpose; millimetres will be found convenient. The wind speed in this case is 20 kt, so that 20 mm should be set on the dividers and transferred to the wind direction line, thus completing the wind velocity.

4. The TAS is 100 kt, so that 100 mm should be set on the dividers. Place one leg on the wind velocity and the other on the track; by joining these two points the triangle of velocities is completed and this is shown in Fig. 19. The last side, true heading (Hdg (T)), is marked ➤—. The Hdg (T) is measured using a protractor at the end of the triangle opposite to the wind velocity.

5. Using the dividers in conjunction with the millimetre scale on a rule, measure the length of the track line to obtain the ground speed (G/S). This process is effected in a matter of seconds with a navigational computer – an essential item for the pilot-navigator.

It should be noted that the three sides of a navigational triangle of velocities represent the following pairs of variables:

1. True Heading and true Airspeed
2. Wind Direction and Wind Speed
3. Track and Ground Speed.

Use of the Navigational Computer

In the preceding paragraphs the triangle of velocity was used to solve navigational problems by plotting to scale the three all-important vectors, viz. track/ground speed, Hdg (T)/true airspeed and wind speed/direction.

The computer eliminates plotting, and when it is used in conjunction with the circular side-rule (usually incorporated in the computer) most navigational problems can be solved easily, quickly and with a high degree of accuracy.

STAGE 1.

Track req. 090°

STAGE 2.

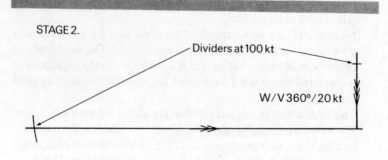

Dividers at 100 kt

W/V 360°/20 kt

STAGE 3.

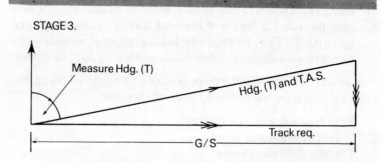

Measure Hdg. (T)

Hdg. (T) and T.A.S.

Track req.

G/S

Fig. 19   Constructing a Triangle of Velocities. Squared paper can be of assistance but it is not essential.

Construction of computer

The Dalton Navigational Computer which became so well known to the wartime service pilot incorporated a moving belt overprinted with drift lines. The instrument was contained within a rather cumbersome metal box and it has now been superseded by the modern lightweight computer which although based upon the Dalton is nevertheless small enough to be carried in a briefcase. Most of these computers can perform two main functions:

1. The solving of triangle of velocity problems
2. Time and distance calculations together with various conversions. These will be explained later in the chapter.

A typical modern computer is illustrated in Fig. 20. It is made of metal or plastic, each side being designed to perform the two functions previously mentioned. Dealing with triangle of velocity problems first, the relevant part of the computer consists of a rotatable compass rose which may be set against a pointer or **True Heading Arrow** at the top of the computer. Attached to the compass rose is a circular transparent plotting area. It has a matt surface so that it can be marked with soft pencil or ink. Ball point pens are to be avoided because they can prove difficult to erase from the surface. Since the plotting area is attached to the compass rose it follows that both components rotate together.

Located in slots within the body of the computer is a movable oblong sheet of plastic overprinted with drift lines or **Fan Lines** which originate from a common point. These are crossed by a series of **Speed Arcs** (also drawn from a common centre) marked in figures from 20 to 250 while the reverse side of the slide is calibrated from 150 to 1050 for high-speed aircraft. These figures may be regarded as kilometres per hour, miles per hour or knots, as required. As this calibrated slide is moved up or down in the computer the markings and figures appear behind the rotatable transparent plotting area.

The only permanent mark on the plotting area is a small central dot. In a serviceable computer this should remain over the centre line of the calibrated slide when the compass rose is rotated. Should it wander off centre accuracy will suffer accordingly.

Solving problems related to Track, Drift, Wind Velocity, Heading and Groundspeed

How to find Hdg (T), G/S and drift.

Assume that a flight is from Leeds-Bradford Airport to Blackpool. The Tr. Req. is 263°, and most economical cruising speed is 120 kt. Cruising altitude is fixed at 3000 ft. The wind velocity at this height is known to be 038°/20 kt. What will be the Hdg (T), G/S and drift?

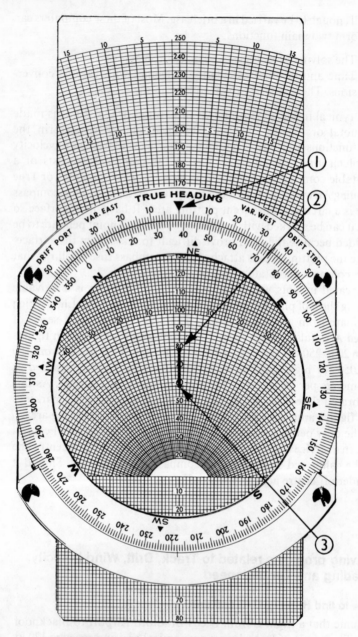

Fig. 20 Finding Heading and Ground Speed: Stage 1.
(1) Set wind direction.
(2) Mark wind speed using sliding scale to measure up from central dot (3).

Method

(*a*) Set the W/V by rotating the compass rose until direction 038° is aligned against the 'true heading' arrow. Using the centre line of the sliding scale as a measuring rule make a small ink or pencil mark on the plotting area 20 units up from the central dot (a line is shown for the purpose of illustration but a dot is more convenient). The wind velocity has now been set on the computer (Fig. 20).

(*b*) Set the Tr. Req. on the computer by rotating the compass rose until 263° is against the true heading arrow. Wind Velocity and Required Track are now set in relation to one another.

(*c*) Set the TAS by moving the sliding scale until the 120 arc is directly under the ink or pencil dot (Fig. 21).

(*d*) To find the Hdg (T) it only remains to determine its angular difference port or starboard of Track. Reference to the sliding scale shows the ink dot to be on the 7° starboard fan line indicating that heading is 7°S of Track.

Surrounding the top portion of the compass rose is a scale marked in degrees port and starboard of the true heading arrow. Transfer the reading from the fan lines to this scale and under the 7° starboard mark read off the true heading, in this case 270°.

(*e*) The central dot on the plotting area indicates the ground speed which is 133 kt.

Note. Finding a Hdg (T) and G/S from a W/V, TAS, and Tr. Req. is the most common requirement of the pilot/navigator. The foregoing procedure represents a major departure from the method usually taught. It eliminates the trial and error procedure used in the older method for finding drift and is in consequence both simpler and quicker.

How to find W/V when Tr. and G/S are known

After making good a Track parallel to the railway which leaves Boston, Lincs, on a bearing of 038° (T), G/S is found to be 100 kt. The Hdg (T) is 048° and TAS is 90 kt. Compute the W/V for the height at which the aircraft is flying.

Method

(*a*) Set Hdg (T) against true heading arrow.

(*b*) Set TAS below dot on plotting dial.

(*c*) Calculate the drift with the aid of the computer or mentally, viz. Tr. = 038° (i.e. parallel to the railway track), Hdg (T) = 048°, therefore Drift = 10°P.

2 Navigation

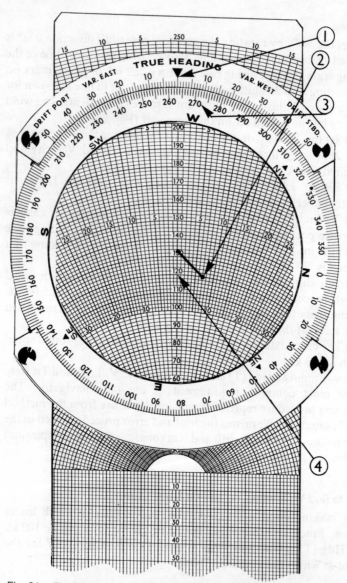

Fig. 21 Finding Heading and Ground Speed: Stage 2.
(1) Set Required Track. Using the speed arcs (4) set TAS under wind velocity mark and determine drift from nearest fan line (2).
(2) Transfer drift to Drift Scale (above the compass ring) and True Heading will be found immediately below (3) in this case 270. Ground Speed (shown under central dot) is 133 kt.

(*d*) Make W/V mark where 10°P drift line intersects the 100 kt ground speed arc. (Fig. 22).

(*e*) Rotate plotting dial until W/V mark is on the centre line in the lower half of the computer.

(*f*) Read wind direction against true heading arrow : 165°.

(*g*) Read wind speed by counting the units, and fraction of units if any, between dot and W/V mark : 20 kt. (Fig. 23).

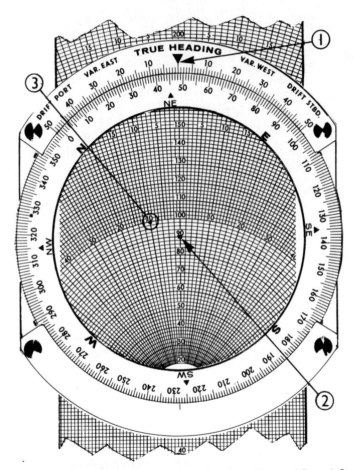

Fig. 22 Finding W/V from a known Track and Ground Speed: Stage 1.
(1) Set Hdg(T) on computer.
(2) Set TAS against central dot.
(3) Place a dot where the known Ground Speed (speed arc) crosses the known drift (fan line).

2 Navigation

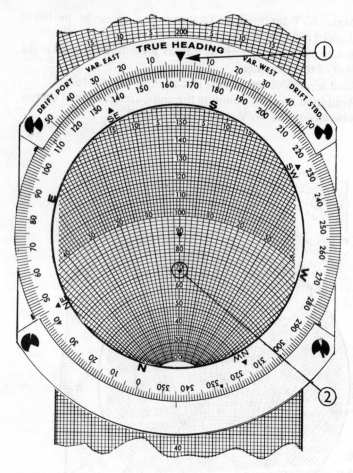

Fig. 23 Finding W/V from a known Track and Ground Speed: Stage 2.
Rotate plotting dial until the dot positioned in stage 1 is on the
centre line below the centre dot.
(1) Read off Wind Direction, 165.
(2) Using the sliding scale, squared area, read Wind Speed
which is 20 kt.

How to find DR Tr. and G/S

For this example assume that the pilot has set heading from Lympne
on 136°(T). TAS is 139 kt and the W/V is 240/30 kt. In order to
determine where he shall cross the French coast he must compute the
DR Track (Dead Reckoning Track). To find the ETA it will also be
necessary to know the G/S.

Method

(*a*) Set W/V by rotating the compass rose until 240° is adjacent to the true heading arrow, then make an ink dot 30 units **down** from the central dot.

(*b*) Set TAS (139 kt) below central dot on plotting dial.

(*c*) Set Hdg (T) (136°) against true heading arrow.

(*d*) Below W/V mark read amount of drift : 11°P. Applied to Hdg (T), this gives a Track of 125°. When plotted on the chart this shows that he should cross the French coast at Ambleteuse, just south of Cape Gris Nez.

(*e*) By reference to the speed arcs a G/S of 149 kt is obtained. As the distance to Ambleteuse is 28 n.m. the aircraft will pass over this place after 11¼ minutes' flying.

Use of the circular slide-rule (Fig. 24)

Attached to the computer, and used in conjunction with it, is the circular slide-rule. With a little practice this, too, will give speedy and accurate solutions to many of the problems associated with air navigation. The following paragraphs explain how to deal with these problems.

Calculating ASI and Altimeter Corrections, Fuel Consumption, Elapsed Time and Converting Units

Changes in temperature affect the density of the atmosphere. This, in turn, will cause inaccuracies in the Airspeed Indicator and Altimeter. Additionally the ASI must be compensated for changes in air density which occur with changes in flight level.

Whereas ASI corrections are important even to relatively low-flying aircraft, altimeter corrections are of more consequence to pilots flying transport aircraft at high cruising levels. Corrections to ASI or altimeter readings are simple to perform on the circular slide-rule which has two windows provided for the purpose.

ASI Corrections

Apart from the effects of temperature and altitude just mentioned the ASI is subject to **Position and Instrument Error**. While every endeavour is made to install the pitot head away from airflow disturbances, errors caused by its positioning do occur, in addition to

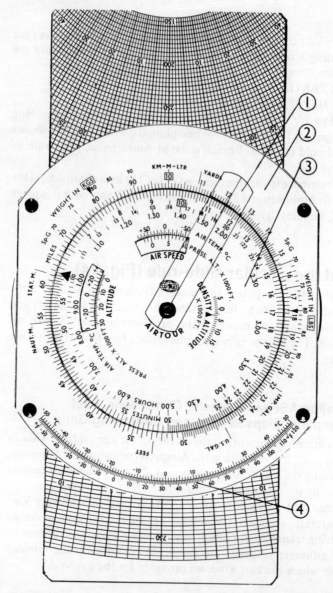

Fig. 24   The Circular Slide-rule. Numbers refer to, (1) cursor used to
line up figures between inner and outer scales: (2) fixed outer
scale: (3) rotating inner scale: (4) temperature conversion
scale. Three small correction windows are also provided: (left)
altimeter corrections: (top) air speed corrections: (right) density
altitude.

inaccuracies in the instrument itself. There are therefore three separate airspeeds and these are listed below:

Name	Abbreviation	Description
Indicated airspeed	IAS	The actual speed indicated on the airspeed indicator.
Rectified airspeed	RAS	This is indicated airspeed corrected for position and instrument error. In the USA this is known as CAS (Corrected Airpeed)
True airspeed	TAS	This is rectified airspeed corrected for height and temperature and it represents the actual speed of the aeroplane in relation to the surrounding air

The correction from IAS to RAS is effected by a correction card which will be adjacent to the airspeed indicator unless the errors are so small that it is deemed unnecessary to have one. RAS is converted to TAS on the navigational computer as follows.

Method

Assuming an indicated airspeed (IAS) of 118 kt and an instrument correction of +2 (found on the calibration card in the aircraft) this would give a rectified airspeed (RAS) of 120 kt. If, in this example the aircraft is cruising at 10,000 ft and the outside air temperature (OAT) reads 0°C, set 0° on the scale at the top of the 'airspeed' window against figure 10 in the window (Fig. 25). The slide-rule is now set to convert any RAS to a TAS for that height and temperature. To convert 120 kt RAS locate figure 12 on the rotatable inner scale and the number adjacent to it on the outer scale is the TAS, in this case 14 or 140 kt – an appreciable difference of 20 kt.

In so far as navigational problems are concerned, all workings must be in TAS. At the height where most light aeroplane flying occurs, the difference between RAS and TAS will vary by 10 kt or so, 5 kt being typical. Conversely a jet aircraft flying at 40,000 ft at a TAS or 500 kt may only indicate 255 kt.

Altimeter Corrections

Indicated altitude may be corrected to true altitude by setting the reading from the altimeter (using the scale along the bottom of the 'altitude' window) against the outside air temperature at that level (figures in the 'altitude' window). Conversion from indicated to true is then made reading from inner to outer scale on the circular slide-rule as with airspeed corrections.

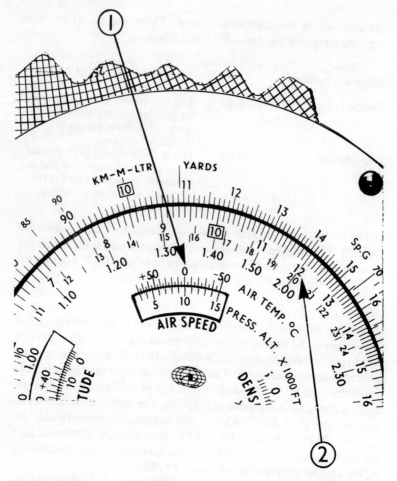

Fig. 25   Finding the True Air Speed.
(1) Using the air speed window set outside air temperature against cruising altitude.
(2) Against Rectified Air Speed on the inner scale read off True Air Speed on the outer scale (in this example slightly more than 140 kt).

For example, the altimeter reads 20,000 ft and the thermometer gives an OAT of –30°C. Using the 'altitude' window, align 20,000 ft with –30°. The circular slide-rule is now set to make the conversion which is completed by locating the indicated altitude on the inner scale and reading true altitude on the outer scale – in this case 19,600 ft.

Calculating Fuel Consumption

At economical cruising RPM a particular aircraft uses 4.3 gallons of fuel per hour. How much petrol would be needed for a flight from Denham to Swansea, a distance of 133 n.m., at a G/S of 93 kt?

Method

(*a*) By calculating as explained under **Elapsed Time** below the flight will take 86 min. To this it is essential to add 20 per cent as a safety precaution against getting lost or having to use an alternative airfield.

(*b*) To find 20 per cent of 86: set 10 (representing 100 per cent) on inner scale against 86 on outer scale. Against 20 on inner scale read off on outer scale 20 per cent of 86 min: 17 min. Now proceed to find the amount of fuel required for a flight of 103 min.

(*c*) Set 60 on inner scale against 4.3 (actually 43) on outer scale.

(*d*) Against 103 on inner scale read off on outer scale fuel needed: 7.4 gallons.

Calculating Time and Distance on Fuel Remaining

Here is another type of problem which may be solved on the computer during a cross-country flight.

The fuel gauge indicates that 3 gallons remain in the tank. How many more minutes' flying will this provide, and how far will the aircraft fly in this time? The fuel consumption and G/S are the same as in the previous example, 4.3 GPH and 93 kt respectively.

Method

(*a*) Set 60 on inner scale against 4.3 (43) on outer scale.

(*b*) Against 3 (30) on outer scale read on inner scale the number of minutes' endurance: 42. Now calculate how far the aircraft will fly in these 42 minuts. Here is the procedure.

(*c*) Set 60 on inner scale against 93 (the ground speed) on outer scale.

(*d*) Against 42 on inner scale read off on outer scale the distance than can be flown on 3 gallons = 65 n.m.

Elapsed Time

On every cross-country flight at least one problem of this nature will occur, e.g. how long (Time) will it take to cover a distance of 225 n.m. at a G/S of 90 kt?

Equationally, this is

$$\frac{\text{Distance}}{\text{G/S}} = \text{Time}$$

$$\therefore \frac{225}{90} = 2\frac{1}{2} \text{ hours}$$

Here is a further example of the same type of problem, one which would take several minutes to solve without the aid of a computer.

From Fair Oaks, in Surrey, to Roborough, near Plymouth, the distance is 145 n.m. How long would this flight take at a G/S of 80 kt.

Method

(*a*) Set 60 (minutes) on the inner scale against 80 (n.m.) on outer scale.

(*b*) Against 145 on outer scale (marked on the computer as figure 14 and 5 small divisions) read on inner scale the number of minutes: 108.

Calculating distance

The simple equation in the previous example now appears in this form

$$\text{G/S} \times \text{Time} = \text{Distance}$$

Here is an example.

After flying for eighteen minutes at a G/S of 90 kt how far has the aircraft travelled along the 48 n.m. Track between Birmingham and Cranfield?

$$\therefore \frac{90}{1} \times \frac{18}{60} = ? \text{ n.m.}$$

Method

(*a*) Set 60 on the inner scale against 90 (n.m.) on the outer scale.

(*b*) Against 18 (minutes) on inner scale read on outer scale the number of nautical miles: 27.

Calculating Speed

To find an actual G/S multiply the distance travelled by 60 and divide by the time taken, in minutes, to cover the distance.

$$\frac{\text{Distance} \times 60}{\text{Time}} = \text{G/S}$$

Here is an example.

Exactly 10½ minutes after setting heading from Hamble, near Southampton, for Dorchester, the pilot pin-points himself over Hurn airfield. The distance travelled is 23½ n.m. What is the G/S?

$$\frac{23\frac{1}{2} \times 60}{10\frac{1}{2}} = ?\ \text{kt}$$

Method
(*a*) Against 23½ on outer scale set 10½ (minutes) on inner scale.
(*b*) Against 60 on inner scale read on G/S outer scale: 134 kt.

Converting Statute Miles, Nautical Miles and Kilometres

There are occasions when it will be necessary to convert statute miles to nautical miles or kilometres, or vice versa. To do this mathematically the following ratios should be memorized:

33 Nautical Miles = 38 Statute Miles = 61 Kilometres
Examples
Convert 88 nautical miles to statute miles.

$$\frac{\overset{8}{\cancel{88}}}{1} \times \frac{38}{\underset{3}{\cancel{33}}} = \frac{304}{3} = 101.5 \text{ statute miles}$$

Convert 114 mph to knots.

$$\frac{\overset{3}{\cancel{114}}}{1} \times \frac{33}{\underset{1}{\cancel{38}}} = 99 \text{ kt}$$

2 Navigation

Convert 133 statute miles to kilometres.

$$\frac{\overset{7}{\cancel{133}}}{1} \times \frac{61}{\underset{2}{\cancel{38}}} = \frac{427}{2} = 213.5 \text{ km}$$

With the circular slide-rule these problems may be solved within a matter of seconds. While computers differ slightly in detail design the examples used to illustrate this book have three marks on the outer scale labelled respectively NAUT.M., STAT.M. and KM-M-LTR (representing kilometres, metres and litres). When converting units of distance markings on the inner or time scale represent nautical miles, statute miles or kilometres as the case may be. By aligning the NAUT.M. mark against any number of nautical miles (inner scale) the equivalent number of statute miles or kilometres can be read off opposite the appropriate mark. The reverse is equally simple. Similar facilities are provided for converting Imperial gallons to US gallons or to Litres.

Here is an example.

How many statute miles are equivalent to 228 km?

Rotate the inner scale until 228 is in line with the KM mark. Against the STAT.M. mark (Fig. 26) read off the equivalent number of statute miles : 142.

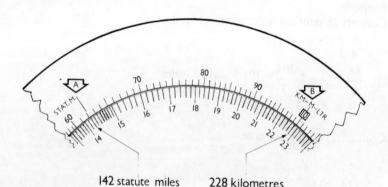

142 statute miles   228 kilometres

Fig. 26   Converting kilometres to statute miles using the kilometre mark (B) and the statute mile mark (A) in conjunction with the inner scale.

Units of measurement

The universal measurement system agreed by the International Civil Aviation Organization (ICAO) is as folows:

Long distances	Nautical miles and tenths
Short distances (runways, cloud height, poor visibility, ground elevation, etc.)	Metres
Altitudes and heights	Feet
Horizontal speed	Knots
Vertical speed	Feet per minute
Visibility	Kilometres or metres
Wind-speed and direction	Degrees and knots
Weight	Kilogrammes
Temperature	Centigrade/Celsius

In-flight Heading Corrections – 1-in-60 Rule

However accurately the heading may have been calculated and however carefully the pilot may steer the aircraft a wind velocity can often prove slightly different to that forecast. Also, winds can change over quite short distances, particularly while flying from coastal regions to a position inland or *vice versa*.

The ability to estimate the number of degrees off track (i.e. **Track Error**) and then determine what heading alteration must be made (*a*) to fly the aircraft back on track, or (*b*) to reach the destination from the present, off-track position, is of obvious value to the pilot. Technique adopted is a simple one known as the **1-in-60 rule**.

The 1-in-60 Rule

Figure 27 shows that an error of 1° will account for an aircraft being 1 mile off track after travelling 60 miles. If the error is not corrected the

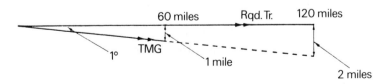

Fig. 27 Basis of the "1 in 60 Rule".

aircraft will be 2 miles away from its Tr. Req. when it has gone twice as far, i.e. 120 miles. Similarly, had the error been 2°, the aircraft would have been 2 miles away from the Tr. Req. after 60 miles or 2 n.m. off track after 60 n.m.; the proportion is 1 in 60 throughout.

The example illustrated in Fig. 28 shows a flight in progress. After flying for 30 n.m. the pilot recognizes that the aircraft is over pinpoint C which is 3 n.m. to the right of track. Using the 1-in-60 rule track error is 6°. The problem now arises, how to regain track?

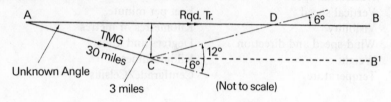

Fig. 28   Application of the 1 in 60 rule after 30 miles.

The error is 6° but an alteration of heading to the left by this number of degrees would merely establish the aircraft on a new track running parallel to the one required and the 3 n.m. track error would remain. Eventually the aircraft would arrive at B¹ instead of B.

Correct procedure is to double the track error (i.e. 12°) and turn 12° to the left. After flying for approximately another 30 n.m. the aircraft will arrive back on track at D. The pilot will then make a 6° right heading alteration to avoid overshooting track and the new heading should fly the aircraft over destination B.

The example in Fig. 29 shows that after covering a distance of 15 miles on a 30-mile flight from X to Y, a pin-point is made at Z, 2 miles to the left of the Tr. Req. What alteration of heading is necessary to reach the destination? Using the 1-in-60 rule, by proportion after 60 miles the aircraft would be 8 miles away from the Tr. Req.; therefore, the track error is 8°. By doubling this amount and altering heading 16° to the right the aircraft will converge on to its destination after flying for approximately the same length of time as it took to reach position Z.

The mathematics of the 1-in-60 rule as they apply to numbers not readily convertible mentally into proportions of 60 are:

$$\frac{\text{distance off track}}{\text{distance flown}} \times 60 = \text{track error}$$

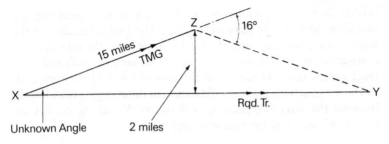

Fig. 29   Application of the 1 in 60 rule after 15 miles.

For example, if after flying 40 n.m. an aircraft is 4 n.m. off track, track error is

$$\frac{4}{40} \times 60 = 6°$$

Applying this to the in-flight situation illustrated in Fig. 30, the aircraft is 3 n.m. to port of track after 30 n.m. flying, with a further 20 n.m. to fly. What heading should be steered to arrive over the destination? By using the 1-in-60 rule the number of degrees to alter heading is easily calculated.

Three n.m. to port after 30 n.m. flying is a track error of 6° P. An alteration of 6° S will make the aeroplane fly more or less parallel to track but 3 n.m. to one side of it. Three nautical miles to port with a

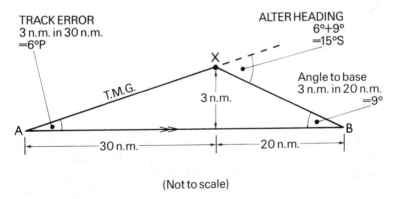

(Not to scale)

Fig. 30   Using the 1 in 60 rule to determine new heading to reach point B from off-track point X.

further 20 n.m. to go amounts to a 9° correction, assuming the aeroplane to be flying parallel to track. The first calculation is the track error from base and the second represents the alteration to destination. By combining the two results, an alteration of heading to reach destination is obtained – in this case 6° + 9° = 15° to starboard.

These corrections make no allowance for the changing relationship between the aircraft's heading and the W/V but the results are sufficiently accurate for practical purposes.

Making good a reciprocal track

Very often a return flight to base will be along the same track as that used on the outward journey. The track direction on return, will be the exact opposite, or, in other words, it will be a **Reciprocal** Track, i.e. different by 180° to that measured on the outward journey.

To make good a reciprocal track a simple rule should be used. It is this:

to the reciprocal Hdg (T), or Hdg (M), apply double the drift in the opposite direction.

For example, on a flight from Oxford (Kidlington) to Bournemouth (Hurn) the Tr. Req. is 198°, Hdg (T) 188° and the drift 10° starboard. What then is the Hdg (T) to make good the reciprocal track of 018° back to Oxford?

Method

The reciprocal Hdg (T) = 188° — 180° = 008°. On the outward journey the drift was to starboard; therefore double the drift (20°) must be *added* to give 10° port drift on the return journey (Fig. 31).

$$008° + 20° = 028°$$
$$\therefore \quad Hdg (T) = 028°$$
$$Drift = 028° — 018° = 10° P$$

The various in-flight corrections just described require of the pilot an ability to map read. Without this skill it will be impossible to recognize ground features on which to fix the aircraft's position. Like all skills, map reading will improve with experience. Meanwhile, the next section explains the basis of map reading and lists the various ground features which are likely to be of value to the pilot.

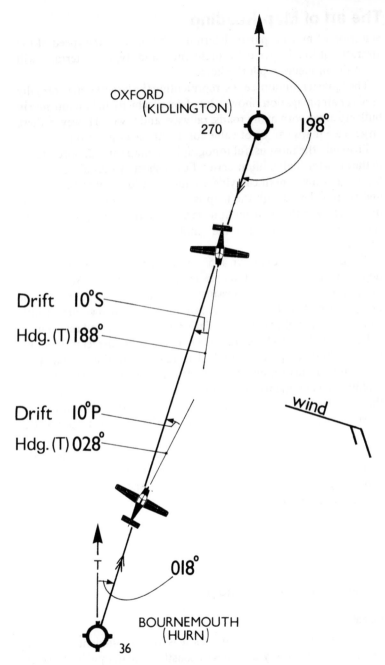

Fig. 31   Making good a reciprocal track.

The art of Map Reading

Selection of maps is partly determined by the cruising speed of the aircraft although personal preference and type of terrain will naturally influence a pilot's choice.

The 'quarter million' series, representing the largest scale normally used for air navigation, shows a lot of detail which can be confusing in built-up areas with many roads, railways and towns. However, these large scale maps can be of value for use at the destination.

Probably the most useful topographical map for light aircraft use in the air is the 'half million' series. Fewer sheets are required to cover flights that may entail carrying a number of 'quarter million' maps and there is less detail; the map is confined to essentials. For long distances even the 'half million' series would entail carrying many sheets and in such cases the 'million' series may prove more convenient.

The use of fan lines drawn at 5° and 10° intervals from the point of departure is explained in Exercise 18, Volume 1 (p. 276). With the aid of these lines an accurate assessment of track error can be made in the early stages of flight. However, fan lines on their own are of little value unless the pilot can recognize ground features and read a map.

For normal flights based upon map reading the correct technique is to read from map to ground. The reason for this is that new, sometimes quite prominent ground features such as reservoirs, large building developments, power lines, etc., could have appeared after the map was printed. On the other hand most of the best features will be shown on the map and these will be sufficient for the purpose of pilot navigation.

Some ground features are of more value to the pilot than others although, with practice, quite insignificant ones may be used to confirm position. So that the various ground features appear in recognizable order it is essential to **Orientate** the map, i.e. hold it so that the drawn track is in line with the direction of flight.

Features to aid Map Reading

Coastline

This is one of the best features as it is very prominent and easily distinguishable. Many stretches of coastline are unique in shape and bearing, thus the likelihood of confusion is lessened.

Water

Lakes, reservoirs, large rivers, estuaries and canals are likewise valuable map reading features. Water, too, stands out quite clearly in moonlight.

Hills and mountains

These can be important landmarks, especially when isolated by lowlands, e.g. Signal Hill near Cape Town, the Wrekin in Shropshire and, on a larger scale, the Blue Mountains, New South Wales. However, lesser hills are not very prominent when seen from above 3000 ft.

Towns

Generally, the smaller the town the easier it is to identify. It should be viewed as a whole and compared with the direction of the rivers, canals, railways and roads which serve it. After the identity of a town has been quite definitely established a particular landmark may help to recognize its identity at some future date, e.g. a castle (Windsor), cathedral (Rheims, France), station, stadium or race course.

Cities

Map reading over, or near, a large city can be most difficult. Railways and roads are so numerous as to be of little use and suburban towns lie so close to the main built-up area that it is impossible to view them in a detached way.

Railways

The greater the number of railways lines which lie beneath Track, the less the pilot should be inclined to trust them. Study the whole visible length of line and its bearing. Observe, too, the junctions, intersections, tunnels, cuttings and embankments. As with all other prominent features, railways should be used not as a separate feature but as a patch in the entire quilt of ground detail.

Roads

Quite often they are of little value in highly populated areas, unless they are unique. Freeways and motorways are very distinctive.

Woods

Many parts of the country are dotted with small woods; this makes for confusion. Added to this, cloud shadows on the ground are often mistaken for woods. Large woods make more reliable pin-points, as they are less numerous and more easily recognizable. However, with extensive areas under reafforestation it is quite possible for a wood to appear on the ground before the maps can be brought up to date. This applies equally as well to the vast number of new housing estates which are continually being built all over the country. Similarly, be wary of uncharted lakes appearing in low-lying areas during rainy periods. To avoid this dilemma of looking on the map for something which is not marked there, read from the *map to the ground* and not from the ground to the map.

Snow

A *light* fall of snow on the ground can often aid map reading, as the useless mass of small detail is blotted out and the main arteries and landmarks are thrown into greater prominence. Heavy snow can obliterate vital ground features.

Incidental aids

When flying over the sea, wind direction below 1500 ft may often be indicated by **Wind Lanes** which are usually evident in all but light winds. Wave motion is *not* a guide to wind direction. Remember, too, the changing position of the sun during the day, e.g. at noon the sun should be on the left wing tip when flying on a westerly heading in the northern hemisphere.

There are times when knowing the mileage to a distant landmark can be of aid to cross-country navigation. Estimating distances accurately from the air comes only with practice, so to gain proficiency it is a good idea initially to ring lightly, on a 1:250,000 map, the home airfield with circles 2 miles apart, up to a limit of about 20 nautical miles radius. Then select prominent landmarks at varying distances from the airfield and note their mileage. In good visibility view these landmarks from different heights above the airfield and use the prepared map to check estimations of the distances. The following table gives distances to the horizon in perfect visbility.

Height of Aircraft	Distance to Horizon
1000 ft	33 n.m.
2000 ft	48 n.m.
3000 ft	58 n.m.
4000 ft	68 n.m.

Flight Planning

Before the start of a cross country flight the following actions must be taken by the pilot:

1. Selection of map(s)
2. Rcute planning with regard to weather, terrain and controlled/restricted airspace
3. Fuel requirements calculated
4. Measurement of tracks, bearings and distances
5. Calculation of headings, elapsed times
6. Preparation of map(s), folding and orientation
7. Preparation of the flight plan and log. Selection of alternates
8. Check weight, balance and field performance
9. Obtain ATC clearance.

These and other related matters are essentially practical and, as such, they are described in Exercise 18, *Pilot Navigation*, (page 269, *Flight Briefing for Pilots, Vol.1.*)

Questions

1 **A line drawn on a topographical map represents:**
   (a) A rhumb line,
   (b) A great circle,
   (c) An isogonal

2 **When measuring required track on a topographical map the protractor should be aligned on a meridian:**
   (a) Near the beginning of the track,
   (b) Near the destination,
   (c) Near the centre of the track.

3 **A track drawn on a Mercator's projection is particularly suitable for:**
   (a) Radio navigation,
   (b) Flight over the Poles,
   (c) Long distance navigation.

4 **One inch on a 1:500,000 scale map represents:**
   (a) Approximately 4 st. m.,
   (b) Approximately 8 st. m.,
   (c) Approximately 8 n. m.

5 **Magnetic variation is:**
   (a) The difference between True and Compass Heading,
   (b) The difference between Magnetic and Compass Heading.
   (c) The difference between True and Magnetic Heading.

6 **When coverting a True Heading to a Magnetic Heading variation is applied by:**
   (a) Adding easterly and subtracting westerly variation,
   (b) Adding westerly and subtracting easterly variation,
   (c) Adding westerly and subtracting easterly variation in the Northern hemisphere and reversing the process south of the Equator.

7 **The required Track is 325°, variation 8°W, deviation 2°E and calculated drift 8°P. What will be the Compass Heading?**
   (a) 339°(C),
   (b) 323°(C),
   (c) 327°(C).

8 **After 40 n. m. the aircraft is over a lake 4 n. m. to port of a track which measures 120 n. m. How many degrees, port or starboard, must the pilot alter heading to fly directly to the destination?**
   (a) 9°P,
   (b) 9°S,
   (c) 12°S.

9 **After ten minutes flying you are over a bridge 25 n. m. along a track measuring 225 n. m. but 3 n. m. to the right. What alteration in heading will be necessary to regain track in ten minutes?**
   (a) 14°P,
   (b) 8°P,
   (c) 14°S.

10 **There are six methods of depicting height above and depth below mean sea-level on topographical and other maps and charts. Can you name them?**

11 **What is an isogonal?**
   (a) A broken line surrounding controlled airspace,
   (b) A line on the compass correction card plotting deviation,
   (c) A line on a map or chart, usually broken, joining places of equal magnetic variation.

12 On a map variation may be shown by the following methods:
    (a) A correction on each meridian and a statement in words,
    (b) Isogonals, diagram, statement in words, compass rose,
    (c) A statement in words accompanied by a corrector box, a
        conversion table in the left margin and Air Information
        Circulars issued from time to time.

13 On the current 1:500,000 series of maps what is the meaning of this
    sign ⊗ ?
    (a) A glider site,
    (b) A naval airfield,
    (c) A disused airfield which may be unfit for use.

14 On the current 1:500,000 series of maps what is the meaning of this
    sign ▲?
    (a) A compulsory reporting point,
    (b) A non-compulsory reporting point,
    (c) An unlighted obstruction.

15 While flying over France your track runs across a spot height
    marked as 600 metres amsl. Assuming you wish to clear the position
    by 1500 ft at what altitude should you fly?
    (a) Approximately 3500 ft,
    (b) Approximately 2100 ft,
    (c) Approximately 3950 ft.

16 You plan to fly from Coventry to overhead the White Horse at
    Marlborough, then return to Coventry. On the outward flight you
    are steering 205° (M) to maintain the track of 188° (T). The
    variation is 8°W. What will be the Heading for the return flight?
    (a) 351°(M),
    (b) 007°(M),
    (c) 035°(M).

17 The difference between TMG and track required is called:
    (a) Wind effect,
    (b) Drift,
    (c) Track error.

18 To assist in estimating TMG in the early stages of a map reading exercise the pilot should employ:
    (a) Five or ten degree fan lines drawn from the point of departure either side of track,
    (b) Time marks,
    (c) Ten-minute marks along track.

19 When lost a pilot should:
    (a) call the destination airfield on the distress frequency.
    (b) make an Urgency call on the frequency in use,
    (c) put out a 'Mayday' call to the Distress and Diversion Service on your existing frequency

20 You are flying to a weekend golf meeting. The aircraft is unable to take the golf clubs and your suitcases in the baggage compartment. You should:
    (a) Tie the suitcases in the baggage compartment and lay the golf clubs lengthwise along the cabin,
    (b) Secure the golf clubs in an upright position between the front and rear seats and put the suitcases in the rear baggage compartment.
    (c) Place the golf clubs in the rear baggage compartment with as much luggage as is permitted and position the remaining suitcase(s) within the cabin.

21 What preparations can be made before or during flight to assist in revising the ETA?
    (a) Divide the track into four equal parts, time the third quarter and revise the ETA when starting the last quarter,
    (b) Obtain a revised W/V over the radio,
    (c) Work out distance flown every ten minutes, using computed ground speed and place marks along the track to represent ten minutes flying.

22 When compiling a flight plan gross errors, e.g. setting the wind or variation in the opposite direction are best avoided by:
    (a) Having the flight plan checked by another pilot,
    (b) Forming the habit of estimating track, drift, distance and time of flight before using the computer,
    (c) Re-checking the Wind Velocity.

A Navigational Computer should be used for the following questions

23 Calculate Magnetic heading from the following data: Track req. 255°; W/V 105/15 kt; TAS 140 kt; variation 8°W.
   (a) 266°(M),
   (b) 244°(M),
   (c) 260°(M).

24 You are planning a flight from Biggin Hill to Leicester East, a distance of 100 st. m. Track required is 333° and the W/V is 290/20 kt. At a TAS of 110 mph the aircraft is known to have a fuel consumption of 5 gal/hr. Allowing 0.6 gal for the take-off and climb how much fuel will be required for the flight?
   (a) 6 gal,
   (b) 5 gal,
   (c) 8 gal.

25 Convert 120 km to nautical miles.
   (a) 75 n. m.,
   (b) 65 n.m.,
   (c) 80 n. m.

26 Your IAS is 140 kt and the correction card for that speed reads +1 kt. The aircraft is at an altitude of 8000 ft where the OAT is +5°C. What is your TAS?
   (a) 124 kt,
   (b) 160 kt,
   (c) 157 kt.

27 Your TMG is 045°(T) and the G/S is 135 kt. The aircraft is on a Compass Heading of 045°, deviation 2°W; variation 8°W, and the TAS is 140 kt. What is the wind velocity?
   (a) 128/25 kt,
   (b) 142/22 kt,
   (c) 322/25 kt.

28 Your TAS is 180 kt, the W/V is 160/20 kt and track 150°(T). What is the True Heading and ground speed?
   (a) 151° and 160 kt,
   (b) 151° and 200 kt,
   (c) 178° and 131 kt.

29 You are flying at an indicated altitude of 10,000 ft. The OAT is
    –10°C. What is your true altitude?
    (a) 10,200 ft,
    (b) 10,550 ft,
    (c) 9800 ft.

30 You are flying a light twin-engine aircraft with a fuel capacity of
    100 gal. The journey is over a distance of 725 n. m., the aircraft uses
    16 gal/hr while crusing at a TAS of 165 kt. On the flight there is an
    average headwind component of 20 kt. Allowing an additional 5 gal
    for the take-off and climb how long will be aircraft be able to hold
    over the destination allowing no reserves of fuel and assuming that
    the same power setting is used? And how far could you divert in that
    time?
    (a) 56 min = 135 n. m.,
    (b) 2 hr = 370 n. m.,
    (c) 75 min = 181 n. m.

31 Your owner's manual quotes the fuel capacity of the aircraft as
    50 U. S. gal. How much fuel in Imperial gallons will it hold?
    (a) 21.6 Imp. gal,
    (b) 41.6 Imp. gal,
    (c) 60 Imp. gal.

32 On a day when the QNH is 1013 mb you are cleared to cruise at
    flight level 75. You plan to climb at a speed of 110 kt when the rate
    of climb will be 875 ft/min and at flight level 75 your TAS is 135 kt.
    How long will it take to arrive over the destination which is
    108 n. m. away, assuming a 15 kt tailwind component?
    (a) 40½ min,
    (b) 42½ min,
    (c) 44 min.

33 Your aircraft is refuelled at an airfield in Spain and you sign for 136
    litres. Convert this to Imperial gallons.
    (a) 30 Imp. gal,
    (b) 37½ Imp. gal,
    (c) 22 Imp. gal.

34 The maximum authorized weight for the aircraft is 3600 lb. You require 50 gal for the flight. With crew and passengers but without fuel the aircraft weighs 3080 lb. How much payload remains for baggage allowing 7.2 lb/gal?
  (a) 120 lb,
  (b) 160 lb,
  (c) 360 lb.

Chapter 3
Meteorology

An understanding of the weather affects many activities but it is of particular importance to pilots, amateur and professional.

The science of meteorology was once regarded as inexact but the advent of computers backed up by satellite photography has greatly improved the accuracy and reliability of forecasting. Also improvements in telecommunications have made weather information more instantly available on a worldwide scale.

There can be no doubt that an understanding of the weather is important to all pilots but whereas several excellent books devoted to the subjects treat meteorology in an academic manner, mainly suited to those entering the profession, this chapter is confined to providing a working knowledge of the principles involved and such topics as practical applications of value to the pilot. The chapter also introduces student pilots to the terminology of meteorology.

In the interests of clarity and to make what could seem a complex subject easier to assimilate, this chapter is divided into three sections:

Section 1: Principles of Meteorology
Section 2: The Weather Forecast
Section 3: The Practical Aspects of Meteorology

Flying accidents that are the result of a mechanical failure are very rare and, thanks to improved standards of flying training, 'pilot error', the term often used by the press to describe an incident following some form of mishandling, is also less common than it used to be. However, it is the **Weather Induced Accident** that continues to figure in the accident statistics in most countries of the world. The problem is not beyond cure; it demands that pilots understand and *respect* the weather. They must also recognize their limitations and those of the aircraft being flown.

SECTION 1: PRINCIPLES OF METEOROLOGY

The Atmosphere

The atmosphere is composed of a number of gases which for practical purposes can be called the air. It is, however, the properties of air, principally temperature, pressure and humidity, which chiefly concern the pilot studying meteorology.

Temperature

Movement of air masses is, on many occasions, a result of the effect of temperature. Heat from the sun causes air to rise and in turn air moves in to replace that which is displaced. Circulation of this kind may be local or it can extend over hundreds, even thousands of miles. Air which is heated sufficiently in this way will continue to rise and expand until it reaches an environment of similar temperature. It is a well-known fact that gases, when compressed, are heated (for example, a bicycle pump becomes hot while pumping). Conversely when gases are allowed to expand there is a temperature drop. Dry air is known as **Unsaturated** and when it is made to rise its temperature decrease with height is at a rate of 3°C per 1000 ft. This is known as the **Dry Adiabatic Lapse Rate**, so named because the method of cooling through expansion is known as **Adiabatic Cooling**.

As the rising air cools, and provided there is sufficient moisture in the air, it will eventually reach its **Dew Point Temperature** and become **Saturated**. Any further cooling will cause condensation to occur and allow the formation of cloud. In the process, **Latent Heat** is released and further temperature falls with altitude will be at a reduced rate which is known as the **Saturated Adiabatic Lapse Rate**. In the lower levels this is approximately 1.5°C per 1000 ft.

The lapse rate figures quoted are average, actual figures being variable according to location. Actual rates of temperature drop with altitude are known as the **Environmental Lapse Rate**, the average value being 2°C per 1000 ft.

Humidity and air masses

The amount of humidity in a particular air mass is to a large extent dependent upon its origin. A mass travelling some thousands of miles mainly over the sea is known as a **Maritime** air mass while a mass conveyed over long tracts of land is known as a **Continental** air mass. Further, the origin and track of these air masses may give another clue to their characteristics, for example Polar Maritime, Polar Continental, Tropical Maritime and Tropical Continental.

Pressure

In the UK pressure is measured in **Millibars** (some countries use the term **Hectopascals** which is another name for the same units). However, the USA continues to use inches of mercury.

Atmospheric pressure is measured at weather stations all over the world. Lines, known as **Isobars** are drawn to join places of equal barometric pressure in the course of preparing surface weather charts. These enable one to recognize areas of low pressure known as **Depressions** or **Cyclones**), areas of high pressure (**Anticyclones**) and boundaries between differing air masses which are known as **Fronts**. These topics are dealt with later in the chapter.

Atmospheric pressure is in a continual state of change and a sequence of charts, plotted at intervals, will show the changes that have taken place in the distribution of the weather system. Pressure changes are of particular relevance to the pilot because they affect the readings of the altimeter in a manner that will later be described.

Clouds

(*Illustrated between pages 100 and 101.*)

Throughout the text clouds are referred to by type name and the following list giving brief descriptions of each classification will help the student to understand the remainder of the chapter. From this list it will be seen that clouds are classified into four main groups according to their height and appearance.

1. High Clouds

Height above 20,000 ft. These clouds are normally composed of ice crystals.

(*a*) *Cirrus* (CI): shapeless, wispy, white clouds. Sometimes referred to as 'mare's-tails'.

(*b*) *Cirrocumulus* (CC): has the appearance of ripples. White in colour and known as a 'mackerel sky'.

(*c*) *Cirrostratus* (CS): a thin, white veil-like cloud. Causes a halo to appear around the sun or the moon.

2. Medium Clouds

Heights between 7000 ft and 20,000 ft.

(a) *Altocumulus* (AC): resembles patches of flattened globular masses sometimes arranged in waves or lanes (also known as mackerel sky).

(b) *Altostratus* (AS): layer of grey-blue cloud of fibrous appearance. May develop to considerable thickness.

3. Low Clouds

Usually less than 7000 ft.

(a) *Stratus* (ST): a dirty grey cloud lying close to the ground.

(b) *Nimbostratus* (NS): a thick dark rain-bearing cloud, often covering high ground. Ragged near rain-depositing areas.

(c) *Stratocumulus* (SC): a layer of patches or rolls of globular clouds, dark grey and light grey in colour. Flying conditions in and below these clouds may be bumpy.

4. Heap Formed Convection Clouds

Heights between 1000 ft. and 36,000 ft. or more.

(a) *Cumulus* (CU): of cauliflower appearance. The entire sky may be dotted with these beautiful crisp clouds, and their formation is often due to convection currents ascending on sunny days. Alternatively cumulus clouds may be formed as a result of frontal activity or **Orographic Uplift**, i.e. when air is forced to rise because of high terrain. They are invariably the product of differential heating. Base of cloud is usually flat. Flying conditions may be turbulent, especially around midday.

(b) *Cumulonimbus* (CB): a cloud of great vertical extent, often seen with well developed mushroom or anvil head which can extend to 35,000 ft (or more in equatorial regions). Associated with electrical storms and lightning. Violent vertical air currents in and near these clouds are hazardous and should be avoided. Heavy icing is also a source of danger and hail discharge from the cloud itself and its anvil top can cause severe damage to the aircraft.

Wind

Wind is air in horizontal motion, caused by changes in temperature between one area and another and the resulting pressure difference.

It is expressed as the direction from which the wind blows in degrees from true north, speed being given in knots. Thus 270/15 would indicate a wind from 270° of 15 kt.

Veering and Backing

A wind is said to **Veer** when there is a clockwise change in direction. A **Backing** wind, on the other hand, changes direction anticlockwise.

Surface Winds

At ground level wind is retarded by contact with the earth's surface and is of lower speed than the free flowing air at higher levels. From this it follows that the wind at 2000 ft, for example, will be greater than that at ground level. As a rough guide the wind at 2000 ft overland is about twice the speed of that at the surface and it veers in direction by some 25°, e.g. surface wind 230/10; 2000 ft wind 255/20. However, the progressive differences or **Wind Gradient** is dependent on a number of factors, not least the wind speed and type of surface over which the wind flows. A rugged surface will cause more friction than the sea. In addition to friction caused by the earth's surface wind encounters considerable resistance in the form of hills, buildings etc. which cause the wind velocity to fluctuate and turbulent, gusty conditions result, becoming stronger as the wind velocity increases. This mechanical form of turbulence makes flying near the ground uncomfortable and it can be dangerous to aircraft approaching to land, at a time when the airspeed is reduced. Landing over hangars, large buildings or indeed in the lee of such structures is never good practice and should be treated with great caution on gusty days.

Land and Sea Breezes

These are examples of local air movements. On a hot day, land near the sea is warmed more quickly than the adjacent water, the temperature of the air over land is increased, it expands and rises drawing in cooler air from over the sea to take its place. As a result, there is an onshore breeze.

At night the direction is reversed as the thermals decline over cooling land. The sea, on the other hand, cools more slowly and eventually air over the sea is warmer and less dense than that over land. Denser air above the land now flows out to sea causing a light off-shore breeze.

Changes of temperature during the day, from maximum to minimum and back, are known as the **Diurnal Variation**.

Fog

Fog is one of the pilot's natural enemies and therefore a good knowledge of the conditions under which it is likely to form is important. It is defined as follows:

> visibility less than 1000 metres: **Fog**
> visibility less than 200 metres: **Thick Fog**
> visibility less than 50 metres: **Dense Fog**.

The formation of fog or cloud is dependent upon the **Relative Humidity** of the air, i.e. the amount of invisible water vapour present in a volume of air compared with the maximum amount of water vapour it could hold at that temperature before becoming saturated (when the water vapour would then become visible). Thus 50 per cent relative humidity means that the air is half saturated; 100 per cent fully saturated. From this information may be found the **Dew-point**, i.e. the temperature at which the air becomes saturated. For example an air mass may be measured as 8°C above its dew-point. Should the temperature drop by more than 8°C the air will become saturated and cloud or fog will form. This is because warm air is able to contain more water vapour than cold and indeed fog is dispersed when the air is heated from above by the sun. Temperature and dew-points are shown in the METAR printout example on page 105.

Radiation Fog

The circumstances under which this type of fog will form occur when there is a clear or clearing sky in the evening, allowing rapid radiation without a blanket of cloud to retain the heat. The cooling that follows causes relatively warm moist air (with a dew-point which is easily reached) to condense forming fog. Calm conditions or very light winds to mix the the lower levels of air would aid the process. It often forms first in river valleys when, at night, cold air flows down hillsides.

Fog is dispersed when the sun raises the temperature of the air above its dew-point. The process may also be aided by (*a*) An increase in wind, and (*b*) The arrival of drier air.

Advection or Sea Fog

This type of fog forms when a moist air mass moves over a cold surface and becomes cooled below its dew-point. It is rather like the behaviour of a cold bathroom mirror when the hot water is being run into the bath. Moisture in the air is cooled by the colder mirror which becomes obscured.

In meteorological terms, the 'mirror' may be a cold sea or snow covered or frozen ground. It often forms overland in association with a thaw of lying snow. **Advection Fog** over the sea is known as **Sea Fog**. It may move over the coast and drift inland with the onset of a sea breeze.

Hill Fog

This is simply low cloud which envelops higher ground. Extreme caution must be exercised when flying in such conditions. Detailed forecasts will denote where the cloud base is likely to cover hills and the surrounding terrain and such areas must be avoided during the flight planning stage and, of course, later while in the air.

Mist and Haze

Mist is defined as the 'obscuration of objects due to water droplets in the atmosphere', usually in conditions of 95 per cent but less than 100 per cent humidity.

Haze is obscuration caused by dry particles, usually diffused smoke or dust. There are no upper limits when either mist or haze visibilities are reported but anything less than 1000 metres is classed as fog.

Precipitation

Rain and Drizzle

Cloud is made up of water droplets, normally only one-hundredth of a millimetre in diameter, which fall relative to the surrounding air. Possibly owing to a process of attraction these particles collide with one another during their descent through the cloud, amalgamating to form larger drops. As the droplets grow in size so their descent velocity increases until they are able to overcome ascending air currents within the cloud and drop to earth in the form of rain. As a general rule the greater the vertical extent of cloud the bigger the raindrop so that layer type clouds produce light rain or drizzle, while heavy rain is associated with cumulus types.

Hail

It has been calculated that an up-current of some 1500 ft/min is required to support a very large raindrop. In large cumulo-nimbus clouds these speeds will very likely be exceeded. The vertical currents tend to fluctuate so that the falling raindrop may be arrested, carried up to the freezing levels of the cloud and dropped several times, on each occasion collecting a further layer of glazed or rime ice until the hail is of sufficient weight to drop free of the cloud. Given the facilities it is possible to cut open a hailstone and count the various layers it has collected while journeying up and down within the cloud. Clearly large hailstones are capable of causing very considerable damage to aircraft encountering them in flight. Since hailstones are associated with the larger cumulonimbus clouds and thunderstorms, they should be avoided by a substantial distance.

Snow and Sleet

Although cloud will remain composed of water droplets at temperatures well below freezing (**Super-cooled**), higher clouds above 20,000 ft or so take the form of ice crystals. However, when the ground temperature drops by more than a few degrees below freezing precipitation is invariably in the form of snow. A close examination under a microscope reveals these to be composed of ice crystals formed in a variety of intricate and beautiful patterns of perfect symmetry, arranged in a basic hexagonal design. Ice crystals readily amalgamate to form snowflakes which will retain their crispness while falling to the ground unless the layers of air beneath the cloud are warm enough partly to melt the crystals when sleet will occur. Sleet or wet snow, i.e. snowfall with temperatures around 0°C, can adhere to aircraft in flight causing problems associated with airframe icing.

Unlike super-cooled water droplets, snow being of little density presents no hazard to flight but the reduction in visibility may be very considerable, even total (white out).

Frontal Systems and the formation of a Depression

The Norwegian meteorologist, Professor Bjerknes is responsible for the now well established explanation of the birth of a depression. In its simplest form this concludes that when two air masses of different characteristics meet they do not readily mix and a boundary, known as a **Front**, exists between them.

As the front moves along the surface of the earth it is replaced by another mass of air which, if warmer, is known as a **Warm Front**. As it moves forward it rises upslope over the colder air forming a wedge and creating a fall in pressure within the wedge, marking the beginning of a depression. The difference between the two air masses on either side of the fronts may only be that of a few degrees in temperature and small changes in pressure and/or humidity. Stages in the formation of a depression are shown in Fig. 32.

Low-pressure systems usually cause bad weather and a deep depression is almost certain to bring low cloud, rain and strong winds. The path of these 'lows' across the British Isles from the Atlantic is usually easterly at a rate of approximately 20 kt.

On a surface weather map (**Synoptic Chart**) a low-pressure system is shown as a series of concentric Isobars, i.e. lines along which the mean sea level (m.s.l.) pressure is of equal value; the lowest pressure will be at the centre and the closer the isobars the greater will be the change in pressure over that area. It is the **Pressure Gradient** which determines the strength of the wind. The advent of satellite pictures enables meteorologists to confirm their synoptic charts with factual evidence. An example is shown in Fig. 33.

Because a fluid will always flow from an area of high pressure to one that is lower it would perhaps be expected that in a low-pressure system, the surrounding relatively high-pressure atmosphere would flow directly in towards the centre. However, as a result of the rotation of the earth, in the northern hemisphere wind at ground level circulates around a depression in an anticlockwise direction, blowing in towards the centre at an angle of approximately 30° to the isobars (Fig. 34). The direction of rotation is reversed south of the equator. At ground level the wind is retarded because of friction caused by hills, trees, buildings, etc. As friction decreases with height so the wind speed increases until at approximately 2000 ft above ground level the forces resulting from the earth's rotation and the low-pressure system become stabilized and the wind blows parallel to the isobars. This is known as the **Geostrophic Wind** (Fig. 35).

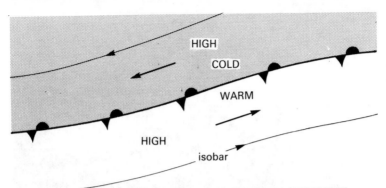

STAGE 1. FRONT SEPARATING COLD AND WARM AIR MASSES

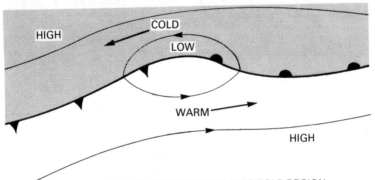

STAGE 2. MOVEMENT OF WARM AIR INTO COLD REGION

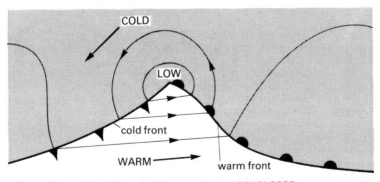

STAGE 3. DEPRESSION FULLY DEVELOPED

Fig. 32 Formation of a Depression:
Stage 1: Front separating cold and warm air masses.
Stage 2: Movement of warm air into cold region.
Stage 3: Depression fully developed.

93

3 Meteorology

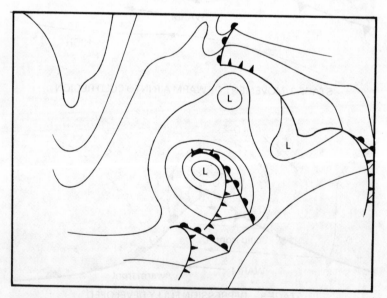

Fig. 33 Satellite picture with synoptic diagram for comparison.
(*Reproduced with the permission of the Controller of Her Majesty's Stationery Office*)

94

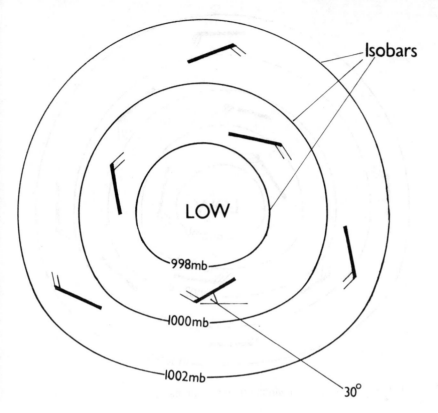

Fig. 34   Flow of surface winds around a Depression.

The Warm Front

When describing the formation of a depression (page 92) it was explained how a mass of warm, moist air rises upslope over an adjacent mass of colder air. Along this gradient a great deal of cloud is formed. The extreme edge of the advancing warm frontal air, up to a height of 30,000 ft or more, is noticeable as wisps of cirrus cloud. To an observer on the ground such cirrus presents an indication that a warm front is approaching. As the front comes nearer the cirrus cloud will be followed by cirrostratus, descending and thickening into altrostratus until finally, reaching almost down to ground level, will come thick, rain-bearing nimbostratus, i.e. stratus of considerable thickness with a low base. Heavy rain or snow from this cloud may extend 150–200 miles ahead of the front and the leading edge of cirrus

95

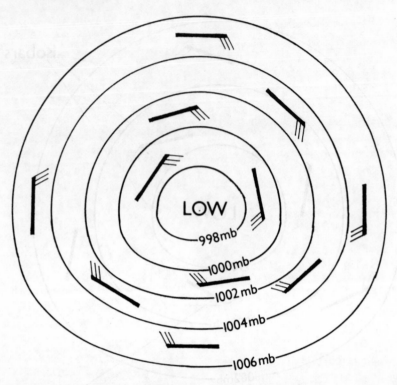

Fig. 35   Flow of upper winds around a Depression.

cloud may be 600 miles ahead of where the warm front reaches the surface. With the passing of the warm front, rain ceases and as the **Warm Sector** arrives the observer will notice a rise in temperature, a change in wind direction (veer) and the barometer, falling during the approach of the front, would either steady or continue to fall at a greatly reduced rate. Flying conditions within the warm sector may include low stratus and stratocumulus with poor visibility and some drizzle. In summer the weather may be pleasant though cloudy. Figure 36 shows a section through the warm front. The warm air rises over the colder air on an incline of approximately 1 in 120.

The Cold Front

Following the warm sector in a low-pressure system is a second front which occurs when colder and denser air intrudes under the relatively warmer content of the warm sector, the frontal incline being steeper, possibly 1 in 50. The warmer air ahead of the advancing wedge of cold air is forced to rise rapidly, creating unstable conditions the more vigorous of which are usually marked by towering cumulus clouds. Some of these may develop into cumulonimbus with thunderstorms and the dangers associated with these conditions. Turbulent conditions and a low cloud base with attendant gusts and squalls are typical characteristics.

A rain belt extending over a depth of some fifty miles marks the passage of a cold front which may be of a more broken character than a warm front. After the warm sector has passed, the observer on the ground would notice a drop in temperature and clearing skies (Fig. 36).

Occlusion

The cold front eventually overtakes the warm front and the two become intermingled squeezing out the warm sector in a process which commences near the centre of the depression gradually extending outwards until no warm sector remains. Many frontal systems moving across the British Isles are not always clearly defined as warm or cold, but as partly occluded. When a cold front overtakes a warm they combine to form an **Occluded Front**, marking the closing stages of the life of the depression. The resulting single front tends to take on the characteristics of a warm or cold front, depending on whether the pre-frontal air was warmer or colder than that which follows, but these characteristics are blurred by the decaying process. A great deal of cloud, rain and drizzle, sometimes prolonged and heavy, may be expected. The various stages of development are shown in Fig. 37.

Secondary Depression

This is a smaller area of low pressure (sometimes referred to as the next developing wave) which develops within the main system, either on a trailing front or within the main system (Fig. 38). Its path around the primary depression is in an anticlockwise direction. Strong winds, rapidly changing in direction, occur under these conditions and are frequently difficult to forecast in the short term without a satellite picture.

3 Meteorology

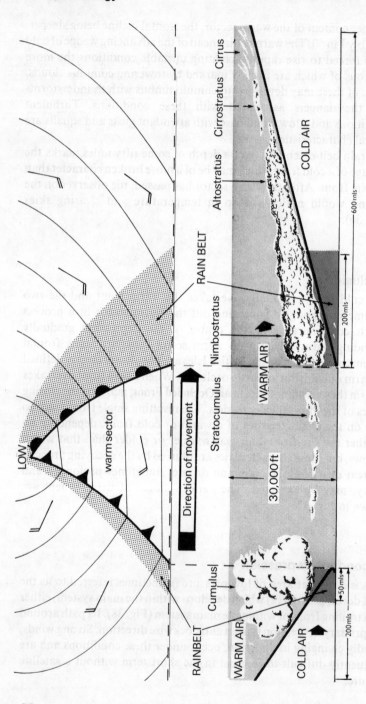

Fig. 36 Frontal systems (not to scale). Illustration compares the synoptic diagram (top) with the weather situation as seen from the ground.

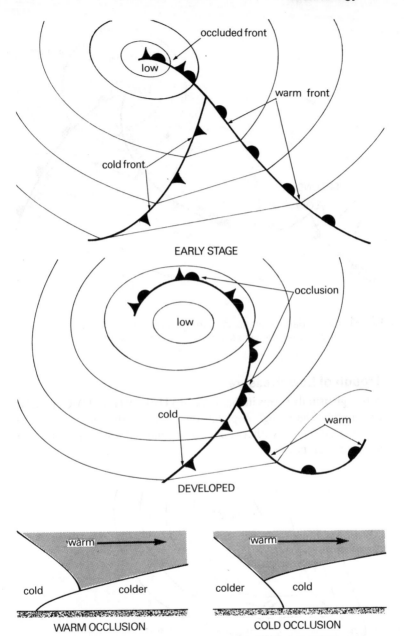

Fig. 37   Formation of an Occlusion. Bottom drawings show two types of occlusing. These depend upon the relative temperatures of the two cold air masses.

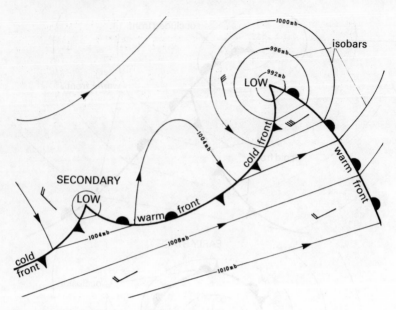

Fig. 38 Secondary Depression forming within the main system.
Sometimes known as the next developing wave.

Trough of Low Pressure

A trough usually forms in the shape of a U or a V (Fig. 39). Generally,
the sharper the trough depicted on the chart the more severe the
weather associated with it and the more marked will be the wind veer
as the trough passes.

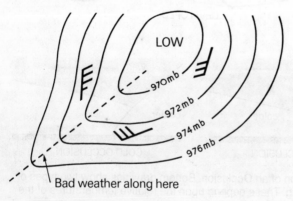

Fig. 39 Trough of low pressure.

Plate 1   Cirrus, with fair weather Cumulus below.
(Courtesy of Kenneth Pilsbury Esq., ISO, FRPS, FRMetS)

Plate 2   Bands of Cirrus increasing to form Cirrostratus (near the horizon), a sign that a warm front is approaching.
(Courtesy of Kenneth Pilsbury Esq., ISO, FRPS, FRMetS)

Plate 3   Cirrostratus with halo. Note the aircraft which is below cloud level.

(Courtesy of Kenneth Pilsbury Esq., ISO, FRPS, FRMetS)

Plate 4   Altocumulus.

(Courtesy of Kenneth Pilsbury Esq., ISO, FRPS, FRMetS)

Plate 5   Clouds typical of an approaching warm front.
(Courtesy of Kenneth Pilsbury Esq., ISO, FRPS, FRMetS)

Plate 6   Stratus forming on hills.
(Courtesy of Kenneth Pilsbury Esq., ISO, FRPS, FRMetS)

Plate 7   Nimbostratus and rain.
(Courtesy of Kenneth Pilsbury Esq., ISO, FRPS, FRMetS)

Plate 8   Stratocumulus.
(Courtesy of Kenneth Pilsbury Esq., ISO, FRPS, FRMetS)

Plate 9   Fair weather Cumulus.
(Courtesy of Kenneth Pilsbury Esq., ISO, FRPS, FRMetS)

Plate 10   Large, growing Cumulus.
(Courtesy of Kenneth Pilsbury Esq., ISO, FRPS, FRMetS)

Plate 11   Cumulonimbus without anvil. Heavy rain is occurring under this thunderstorm.
(Courtesy of Kenneth Pilsbury Esq., ISO, FRPS, FRMetS)

Plate 12　Thunderstorms on a vigorous cold front.
(Courtesy of Kenneth Pilsbury Esq., ISO, FRPS, FRMetS)

Plate 13   Heavy thunderstorm. Note the well developed anvil on the
large Cumulonimbus. Hail is falling on the right.
(Courtesy of Kenneth Pilsbury Esq.. ISO. FRPS. FRMetS)

Plate 14   Mountain wave clouds, wind blowing from right to left (see
Fig. 51).
(Courtesy of Kenneth Pilsbury Esq., ISO, FRPS, FRMetS)

Anticyclone or High

Anticyclones are associated with fair or fine weather and usually move quite slowly across the country, often becoming stationary for several days. Fig. 40 shows how a high appears on a synoptic chart. The arrows indicate light winds circulating in a clockwise direction and deflected outwards at ground level. At 2000 ft the direction of the wind is along the isobars (in the southern hemisphere winds flow round a high in an anticlockwise direction).

Ridge of High Pressure

These bring fine weather of from 12 to 24 hours' duration followed by deteriorating conditions. As the ridge approaches, the wind veers, slackens, and cloud decreases; as it passes, pressure begins to drop, the wind backs and usually increases, and cloud thickens. This is typical of a frontal ridge, i.e. a ridge between two fronts.

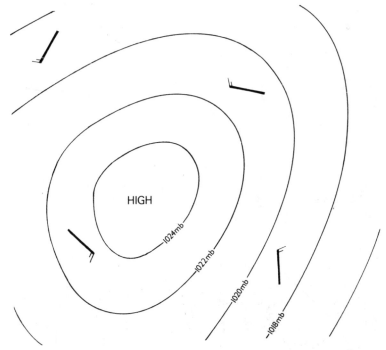

Fig. 40   High pressure system showing the wind flow at ground level. Upper winds circulate in a direction parallel to the isobars.

Col

The slack pressure region between two low or two high pressure areas
is known as a **Col**. During summer conditions overland are likely to
be unstable giving rise to thunderstorms during the afternoon and
evening.

In winter the stagnant conditions may give persistent frost and fog
(Fig. 41).

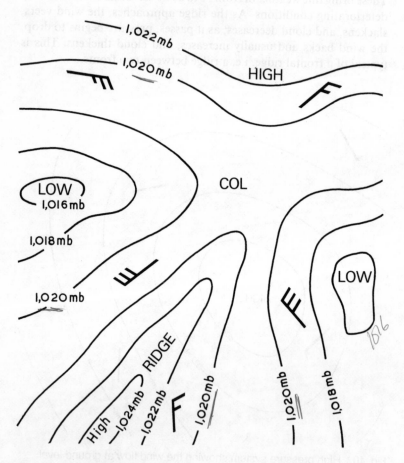

Fig. 41 A Col formed between two low and two high pressure regions.

SECTION 2: THE WEATHER FORECAST

Weather information, that is to say, temperature, humidity, barometric pressure, cloud type and amount, etc., is collected by thousands of meteorological stations situated all over the world both on land and at sea. In most countries this information is fed through a teleprinted network to a Central Meteorological Office and then circulated to the outstations. The reports are conveyed by an international code; data from each station take the form of groups of numbers and letters which give details of weather elements.

The position of each reporting station is marked on purpose-designed weather maps and when the information has been transcribed by the receiving office it is added in symbol form around the reporting station to which it refers. The positioning of the symbols around the station marked on the weather map is in standard form and the complete unit is called the **Station Circle** (Fig. 42).

When all the station circles have been plotted on the weather map the forecaster will draw in isobars joining places of equal barometric pressure (see **Pressure**, page 86). From the completed synoptic chart the weather prevailing at each reporting station can be recognized. Furthermore, by comparing the current chart with one drawn several hours previously a weather pattern will emerge with sufficient clarity to enable forecasts to be made. An example of a synoptic chart is shown in Fig. 43.

Area Forecasts

For meteorological purposes the British Isles and surrounding seas are divided into a number of areas for which a forecast is prepared by the parent meteorological station in that area. Based upon the general inference local geographical features are taken into account to arrive at a detailed forecast within each area.

METAR Code

Of considerable relevance to the pilot is the coded weather information in Meteorological Aerodrome Reports. An example of teleprinted weather information is given overleaf:

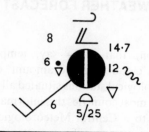

Example Plotted	Met. Element Represented	Meaning of Example
⌐⌐⌐	Type of high cloud	Cirrus
⫽	Type of medium cloud	Altostratus
14·7	Pressure reduced to m.s.l.	1,014·7 mb
12	Barometric tendency	Barometer has fallen 1·2 mb since previous reading
∿	Barometric tendency	Falling unsteadily
▽	Past weather	Showers
⌒	Type of low cloud	Cumulus
5/25	Amount and height of low cloud	Five-eighths of sky covered at 2,500 ft
6	Dew-point temperature	6°C
✓	Wind-speed and direction	215°/13–18 m.p.h.
▽̇	Present weather	Rain showers, slight
6	Visibility	6 kilometres
8	Temperature	8°C
◉	Circle indicates position of reporting station: total cloud amount plotted within circle, e.g. ⅞	

Fig. 42   The Station Circle. Symbols in this example are explained in the table, clockwise from the top.

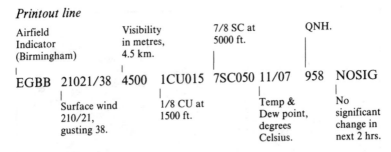

Printout line

Airfield Indicator (Birmingham)	Visibility in metres, 4.5 km.		7/8 SC at 5000 ft.	QNH.	
EGBB 21021/38	4500	1CU015	7SC050	11/07 958	NOSIG
Surface wind 210/21, gusting 38.		1/8 CU at 1500 ft.	Temp & Dew point, degrees Celsius.		No significant change in next 2 hrs.

While this information is available to a pilot at meteorological offices, it can also be obtained over the telephone in plain language.

AIRMET Service

This service is available to pilots over the telephone or via telex. It covers the UK up to 15,000 ft (winds to 18,000 ft) and enables them to determine if the weather conditions are consistent with VFR or IFR. Forecasts are updated four times daily and if necessary amended more frequently. AIRMET forecasts covering the period 0545 hrs to 2300 hrs are recorded. Those for the period 2000 hrs to 0600 hrs will be given by a met officer.

Information should be taken down on **Pilot's Proforma** CA1701 (Fig. 44). The reverse side (Fig. 45) shows that the service is divided into three **Regions** containing smaller, numbered **Areas**. When flying from, for example, SOUTHERN ENGLAND to a destination in NORTHERN ENGLAND AND WALES it would be necessary to obtain met information from each Region. Note also the separate availability of TAFs (page 109) and METAIRs (page 103).

A familiarisation call on the telephone number related to the Region required will demonstrate the precise form and content of the recorded forecasts. Pilots requiring clarification of any particular aspect of a forecast may obtain further information by telephoning one of the meteorological offices providing a service to General Aviation. These are listed in CAP32 – MET.

AIRMET does not replace the principle method of pre-flight met briefing and the use of weather charts but it is an alternative service for the areas covered.

Forecast Weather below 15,000 ft. (LAO Form 215)

The forecast information that is readily available to pilots will depend on the facilities at their aerodrome.

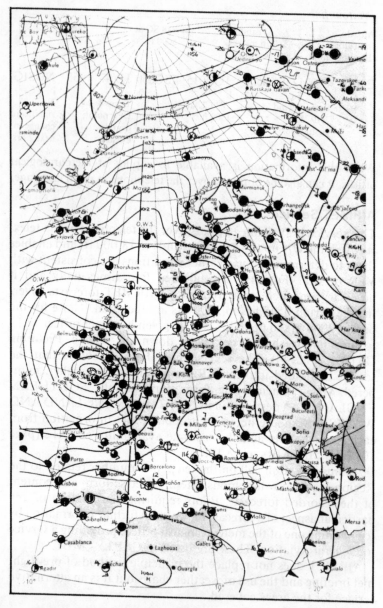

Fig. 43 A synoptic Chart covering Europe, Iceland, part of Greenland and North Africa. (*Courtesy The National Meteorological Library*).

UK AIP (2 Jul 87) MET 2-7

Civil Aviation Authority

AIRMET

PILOTS PROFORMA

NOTES 1 Forecast area coverage and telephone numbers are shown on the map on the reverse.

2 All altitudes are above Mean Sea Level, all times are UTC.

FORECAST FOR: Southern England/Northern England/Scotland *(Circle appropriate area)* **PERIOD OF VALIDITY:** From .... hours to .... hours on .... / .... / .... *(insert times and day, and add date)* **METEOROLOGICAL SITUATION:**		
WEATHER		
WARNINGS AND REMARKS		
AREAS **WINDS & TEMPERATURES** (Degrees True/kt/°C) 2000 feet 5000 feet 10 000 feet 18 000 feet		
AREAS **CLOUD** (Up to 15 000 feet or higher tops if significant to aviation)		
AREAS **SURFACE VISIBILITY** (In metres to 5000 m, km above 5000 m)		
FREEZING LEVEL (0°C Isotherm)		
ICING (Airframe)		
OUTLOOK TO: .... hours		

CA 1701
250287

Fig. 44 Pilot's Proforma for recording AIRMET forecasts.

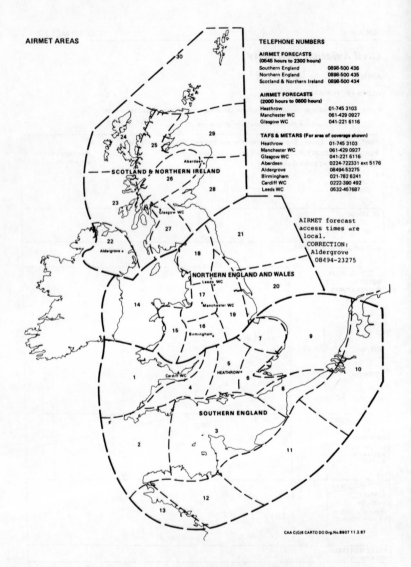

Fig. 45   AIRMET Regions, Areas and telephone numbers.

In the UK low level (i.e. below 15,000 ft) area and route forecasts are issued by the Met. Office, London (Heathrow) Airport. These are transmitted to aerodromes suitably equipped to receive them. The information and weather charts are issued from automatic printout machines and examples are shown in Figs. 46 and 47. Explanatory notes are available (reproduced in Appendix III) and these enable the pilot easily to understand the information issued. When additional information or clarification is required this may be obtained from the nearest met. office which will usually be situated at a major airport.

Terminal Aerodrome Forecasts (TAF)

Of particular importance to the pilot is the forecast weather at the destination and alternative airfields. These forecasts are produced in code form similar to METARs (p. 105) which are actual reports whereas TAFs are forecasts, i.e. conditions expected in the future. An example of a TAF and its decode is shown in Fig. 48. They are also available in plain language from a met. office.

Route Forecasts

In addition to the service already mentioned a weather forecast may be prepared for a specific route and time period and should be requested a minimum of 2 hours before ETD. During the flight the pilot should relate the forecast weather to that experienced and make use of Volmet on the appropriate VHF frequency. If there is any significant unforecast deterioration the pilot must at that stage decide whether to continue the flight to the destination field.

VOLMET (Routine Reports)

Obtainable on designated RTF frequencies on a continuous basis VOLMET provides terminal airfield information in the following order:
Time of observation
Name of terminal (airfield)
Surface wind
Ground visibility
Weather
RVR (Runway Visual Range) if visibility is below 1500 metres

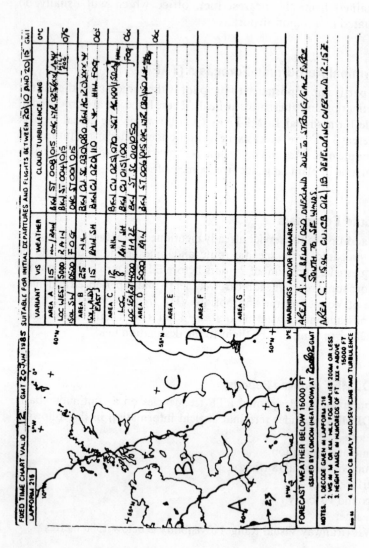

Fig. 46 Typical low level weather forecast issued at airfields equipped with automatic printout machines.

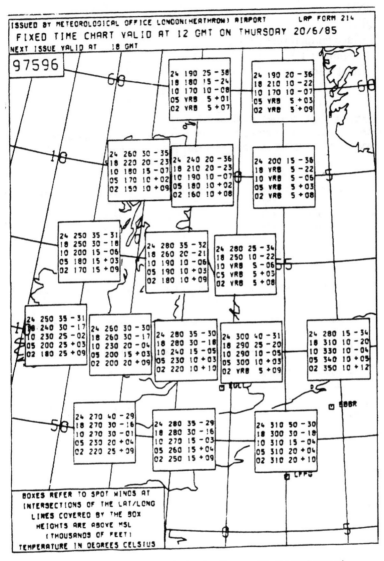

Fig. 47   Automatic printout of wind velocities and temperatures at
altitudes above msl. of 2000,5000,10000,18000 and 24000 feet
shown on an area basis.

3 Meteorology

Cloud
QNH
QFE
Surface temperature (°C)
Dew-point (°C)
Special warnings

<u>Example of TAF</u> (Terminal Aerodrome Forecast)

EGBB 0918 27010 3000 11MIFG 5ST005 GRADU 0911 9999 6ORA 4CU020 INTER 1218
5000 81XXSH 6CU015 PROB 30 TEMPO 1518 1200 96TSGR 7CB010 =

<u>Decode</u>

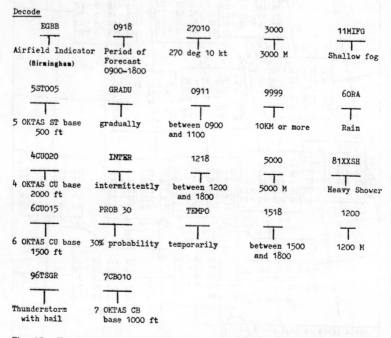

EGBB	0918	27010	3000	11MIFG
Airfield Indicator (Birmingham)	Period of Forecast 0900–1800	270 deg 10 kt	3000 M	Shallow fog

5ST005	GRADU	0911	9999	6ORA
5 OKTAS ST base 500 ft	gradually	between 0900 and 1100	10KM or more	Rain

4CU020	INTER	1218	5000	81XXSH
4 OKTAS CU base 2000 ft	intermittently	between 1200 and 1800	5000 M	Heavy Shower

6CU015	PROB 30	TEMPO	1518	1200
6 OKTAS CU base 1500 ft	30% probability	temporarily	between 1500 and 1800	1200 M

96TSGR	7CB010
Thunderstorm with hail	7 OKTAS CB base 1000 ft

Fig. 48   Example of a Terminal Aerodrome Forecast (TAF) explained on an item by item basis.

Met. Reports

Significant Met. (SIGMET)

These reports are issued where one or more of the following are expected to occur or have been experienced and reported by aircraft:

Active thunderstorms
Tropical revolving storms
Severe line squalls
Heavy hail
Severe turbulence
Severe airframe icing
Marked mountain waves (i.e. vertical currents exceeding 500 ft/min)
Volcanic ash

Special Aerodrome Reports (SPEC)

Whenever there are modifications to Routine Reports (e.g. onset or clearing of freezing rain, moderate or heavy snow, hail, thunderstorms, significant changes in cloud base and/or visibility, etc.) a **Special Report** is issued to the appropriate Air Traffic Service.

Summary

It will not be necessary to use all the forecasting and reporting facilities described in this section and their selection will, of course, be related to the proposed flight and the availability of met. information at the local airfield.

It is always of value to obtain met. information on AIRMET before leaving for the airfield. In addition there are various services available via television for which a fee is charged (e.g. PRESTEL).

Whatever the source of information it is essential to ensure that weather reports and forecasts are current. There is little point in starting a flight on the basis of met. information that is many hours old. Indeed such a practice could be potentially dangerous.

SECTION 3: THE PRACTICAL ASPECTS OF METEOROLOGY

A wide knowledge of meteorology is of obvious importance to the pilot and while some aspects may be theoretical or academic, others of a more practical nature directly affect the flight of the aircraft, the accuracy of its instruments and the behaviour of airframe and engine. This section of the chapter deals with types of weather likely to affect the safety of the aircraft.

Ice Accretion

In the atmosphere water exists in three forms: gas (water vapour), liquid (water droplets) and solid states (ice). It is often erroneously supposed that airframe icing can only occur in cloud but it is more prudent to assume that icing may develop at any time when the temperature is between 0°C and –40°C and there is sufficient humidity. Airframe icing can occur in several forms.

Hoar Frost

Although this is unlikely to form during flight at cruising level it may affect an aircraft when its external surfaces have been cooled at altitude. During a descent, should warmer moist air be encountered, hoar frost may develop, possibly on the windscreen. When hoar frost forms on an aircraft left in the open overnight it MUST be removed prior to take-off since it has an adverse effect on field performance, even to the extent of preventing the aircraft from taking off or climbing.

Rime Ice

This is likely to build up on the leading edges of the wings, tailplane and fin. It takes the form of irregular deposits that look not unlike sugar icing.

Glazed or Clear Ice

Representing a most serious hazard to flight, this type of icing may build up very rapidly in temperatures just below 0°C, materially altering the contours of the leading edge of the wing with attendant aerodynamic deterioration (Fig. 49). When icing of this kind occurs it is important to change flight level by climbing (engine power permitting) or descending out of the icing temperature range (with due regard to safety heights and air traffic considerations). Aircraft experiencing such conditions will invariably request ATC to allow a change of altitude/flight level when circumstances demand.

Principal cause of Glazed Ice

It is common knowledge that at temperatures below freezing water turns into ice but in the atmosphere very pure water can remain fluid

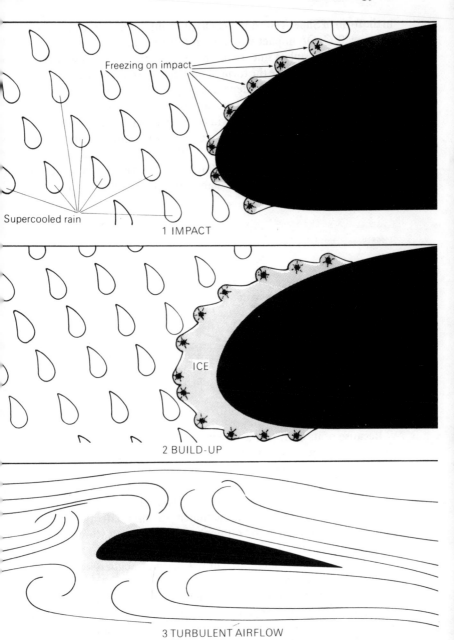

Fig. 49   Formation of Glazed Ice. Smearing and freezing action of
super-cooled raindrops eventually results in a build-up (Stage 3)
which destroys airflow over the wings.

115

in the form of **Super-cooled Water Droplets** although their temperature is below 0°C. On impact with an aircraft their state is changed; the super-cooled droplets smear and freeze on to the contacting surface. When the droplets take the form of rain then the term **Freezing Rain** is used. Rate of build-up is dependent upon the amount and size of water droplets and their temperature, etc. The degree of icing which may be expected during flight will be given in the forecast as 'slight', 'moderate' or 'severe' (See page 349 for symbols). It is the responsibility of the pilot to be aware of the conditions under which icing can occur. The forecast will include a **Freezing Level** (i.e. 0°C). This changes when passing through a frontal system (Fig. 50).

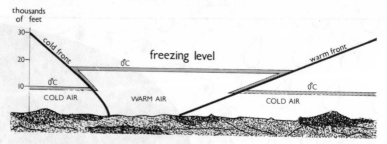

Fig. 50   Changes in Freezing Level within a frontal system.

Because of its detrimental effect on the aerodynamic efficiency of the wing, icing will raise the stalling speed, reduce the effectiveness of the controls and decrease the airspeed and climb performance while at the same time increasing the weight of the aircraft. Modern **De-icing** and **Anti-icing** equipment, which may operate by thermal, fluid or pneumatic means, has done much to eliminate this hazard although it is not fitted to training aircraft. There is, however, the additional consideration of keeping aerials, insulators and antennae free of ice, but modern equipment has design safeguards. The pitot head is electrically heated.

In addition to airframe icing the engine must also be protected. For example a build-up of ice in the air intake will reduce the volume of air reaching the power unit and in piston engines the mixture will be seriously affected. Ice will readily form within the carburettor even when the outside air temperature is well above freezing, particularly in moist air (page 227).

Heated intakes and hot air ducting have largely eliminated these problems, but the pilot must nevertheless be aware of the possible dangers and fully understand the operation of the carburettor heat control.

Propellers are by no means immune from icing which may cause vibration. Furthermore, ice shedding from a propeller, apart from causing unexpected and alarming noises if it strikes the fuselage, may be sufficient to cause damage.

Turbulence

Although clouds are one of the principal sources of turbulence, in windy conditions land features will create disturbances of which the downdraught is an example. Downdraughts can be very severe particularly on the leeward side of hills, and powerful enough to prevent an aircraft of low power climbing over the high ground. This should always be borne in mind when airfields are situated near features likely to cause air disturbances of this kind.

The behaviour of wind near uneven ground may be illustrated by watching a fast-flowing stream as it makes its way over a rock-strewn bed. To overcome a stone the water must rise in a small wave and then descend leaving behind an area of foam and eddies. If the obstacle in the stream is large enough, the flow may even reverse, encouraging flotsam to move upstream. Air in motion near the ground will behave in a very similar manner. In addition to the horizontal flow there are up-currents, down-currents, eddies and gusts all of which contribute to a fluctuation in surface wind velocity. Hills, ridges and even large buildings all create ground turbulence and both the windward and leeward sides of such obstacles may present a danger. Fig. 51 shows the flow of air over a high ridge and the path of an aircraft attempting to fly overhead. The pilot approaches the summit with what he considers to be a safe margin of height. On entering the area of downdraught the aircraft descends and if it is low-powered the pilot may be unable to maintain height or have sufficient speed or room to turn away from the danger. This potential hazard must be understood, and even relatively small irregularities in terrain may give rise to such conditions. In certain meteorological conditions typically with the wind blowing more or less at right angles to the ridge or hills the resulting descending air on the leeward side can be a serious hazard even when smooth conditions prevail and may continue as a **Wave Cycle** (Fig. 51). Wave lengths are dependent on

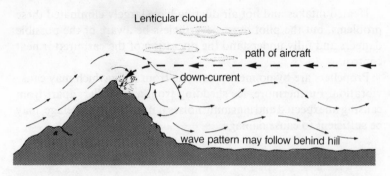

Fig. 51 Dangers of a down-current in the lee of a hill. Presence of cloud in the position shown is a clear warning to pilots flying in mountainous regions.

temperature, stability and wind velocity but 4–5 n.m. is commonplace. The waves can produce local turbulent gusty conditions which do not exist elsewhere and are difficult to forecast.

It is therefore most important to clear hills by a margin of several hundred feet when the wind is more than 15 kt or so, remembering that the wind in flight will be stronger than the surface wind experienced during take off. Good terrain clearance and adequate engine power are excellent safeguards.

During summer small cumulus clouds (sometimes called **Fair Weather Cumulus**) will indicate the existence of thermal or warm air currents which are the result of uneven heating of the ground. For example, air flowing over sand, large buildings or asphalt areas will take on a higher temperature than air over green fields or water. Air warmed by these more radiant areas will rise, its place being taken by a flow of cooler air from surrounding regions and thermals will occur giving rise to more pronounced cumulus cloud development.

Occasionally thermal turbulence can be very active causing flying at lower altitudes to be unpleasant. Rare cases have been reported where bumpiness has been sufficient to cause injuries to the occupants of even large aircraft.

Thunderstorms

Flying through or even close to a cumulonimbus may only be described as foolhardy. However, the development of **Weather Radar** and radio aids to navigation enables these storms to be avoided wherever possible. Weather radar is carried in the nose of many large or high-flying aircraft so that cumulonimbus clouds may be detected well ahead and, in the interest of passenger comfort, avoided, but there are times when for flight planning, air traffic or other reasons, it is necessary to fly through thunderstorm areas.

The problems associated with turbulence and icing have already been mentioned. It is also not uncommon to experience a good deal of hail when entering the storm. Lightning itself, although distracting, is not of very great importance and even a lightning strike may do little more than affect the compass, interfere with radio reception and produce a smell of ozone. All components of the aircraft are **Bonded**, i.e. connected together in electrical continuity so that during a thunderstorm, electrical discharge is able to spread evenly without 'flashover' between components, thus preventing serious damage to the aircraft or harmful effects to the occupants. Circumstances permitting, the worst turbulence can often be avoided by penetrating the storm at flight levels below 10,000 ft.

Sometimes cumulonimbus may be embedded within other types of cloud, making them more difficult to detect from a distance. In such cases the code EMBD will appear in the forecast. TS indicates thunderstorms.

When flight through a thunderstorm or any heap cloud is imminent and unavoidable the pilot should take the following actions:

Before entry

1. Instruct the passengers to fasten their seat belts securely.
2. Secure all loose articles.
3. Check the pitot head is switched on.
4. Disengage the autopilot (if fitted).
5. Turn up the cockpit lighting to its maximum brightness. This will prevent temporary blindness in the event of lightning flashes.
6. Switch off any radio equipment not required.
7. Operate the anti-icing and/or de-icing equipment if fitted.
8. Reduce power to the recommended setting for flight within turbulence and re-trim to hold the lower airspeed that will result.
9. Check the vacuum and/or electric supply to the instruments.
10. Before entering the storm synchronize the direction indicator with the magnetic compass.

After entry

1. Concentrate on controlling the aircraft with special reference to
    lateral and fore and aft level (attitude indicator). Avoid any
    control movements likely to add to the severe airframe stresses
    set up by the storm.
2. Do not overcorrect the fluctuating readings of the altimeter and
    the airspeed indicator.
3. Where possible, avoid turns other than for small heading
    corrections.

After flying through an electric storm the magnetic compass can be
affected and a compass swing may subsequently be necessary.

From the foregoing it will be apparent that pilots of light aircraft
should keep well clear of these clouds.

The Altimeter

Because the altimeter is affected by weather changes its use is
included in this chapter. The average mean sea level barometric
pressure is regarded as 14.7 lb/sq in., (1013.2 m.b.), i.e. the weight of
air bearing on each square inch of surface area. A mountain top
10,000 ft above sea level is subjected to a column of air 10,000 ft
shorter than one reaching down to sea level and being shorter it will
naturally exert less pressure than the sea level value of 14.7 lb/sq in. It
therefore follows that atmospheric pressure decreases with increase
in height above sea level and upon this fact is founded the concept of
the pressure altimeter. Basically the altimeter is a very sensitive
aneroid barometer with a dial calibrated in feet or metres instead of
pressure units. The instrument is calibrated to International
Standard Atmosphere conditions which assume a sea-level pressure
of 1013.2 m.b., a sea-level temperature of 15°C and a decrease in
temperature with height at a rate of 2°C/1000 ft up to a height of
36,000 ft. Temperature is assumed to be constant above this height.
Barometric pressure alters from day to day and area to area so that
provision is made for setting the altimeter by means of a small knob at
the bottom of the instrument. This is adjusted in conjunction with a
Sub-Scale on the face of the instrument, the scale being calibrated
either in millibars or inches of mercury.

Altimeter Settings

Air Traffic Control will give the pilot an altimeter setting (in millibars
or inches or mercury as required) which can take the following forms:

1. Airfield QFE. With this setting the altimeter will read zero when the aircraft is on the ground regardless of the height of the airfield above sea level. This setting enables the pilot to know how far he must descend before touch-down. On the QFE setting vertical distance above the ground is reported as a **Height**.

2. Airfield and Regional QNH. Barometric pressure may vary quite considerably from one area to another particularly when a depression with steep pressure gradients covers the country. The UK and surrounding seas are divided into designated **Altimeter Setting Regions** (Fig. 52) and a setting for any of these areas is always obtainable from the controlling authority. When a Regional QNH is given this will be the lowest forecast value for the next two hours for the region and with this setting the altimeter will read **Altitude** above sea level enabling the pilot to assess his terrain clearance from the map when in proximity to high ground. When flying from one altimeter setting region to another the pilot must obtain the relevant Regional QNH. When the airfield QNH is set the altimeter will read airfield elevation above mean sea level (a.m.s.l.) if the aircraft is on the ground.

3. Standard Altimeter Setting. At the beginning and end of a flight when the aircraft is taking off or landing, height relative to airfield level is all-important (QFE setting). During the climb away from or descent towards the airfield it is essential that the pilot should be aware of his proximity to hills, television masts or other high obstructions. The height of these obstructions is shown on maps as an 'above mean sea level' figure so that during the initial climb out the pilot will change his altimeter setting from QFE to QNH, the instrument then reading altitudes above sea level. At a prescribed altitude (which is usually 3000 ft a.m.s.l. at airfields outside controlled airspace within the UK), the aircraft is said to have reached **Transition Altitude** above which the standard altimeter setting of 1013.2 m.b. (29.92 ins) will be used. This means that above transition altitude all aircraft fly on the same altimeter setting regardless of the current barometric pressure. The obvious advantage of this arrangement is complete uniformity of all altimeter settings and therefore more precise separation when two or more aircraft are flying on instruments over the same area with only the altimeter to prevent the possibility of collision. When the altimeter is adjusted to the standard altimeter setting vertical distance from that datum is given in **Flight Levels:** thus 4000 ft is referred to as flight level 40 and

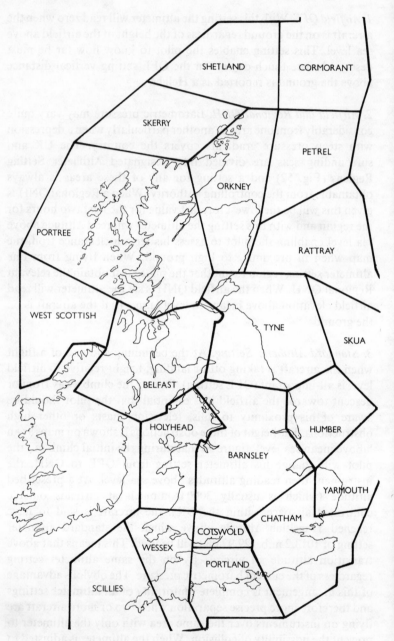

Fig. 52   Altimeter Setting Regions covering the UK.

flight level 150 means that the altimeter will read 15,000 ft. It should be noted that, say, flight level 80 will in fact be 8000 ft above sea level only when the m.s.l. barometric pressure is 1013.2 m.b. As a result of pressure changes it could be 8300 ft one day and 7900 ft the next, but since all aircraft fly above transition altitude with altimeter at the same setting this is of no consequence unless a mountain range has to be crossed while flying on instruments, an aspect which is explained in the next section.

During the descent at the end of the cruise the Flight Level at which the altimeter is changed from the Standard Setting to QNH is known as **Transition Level** and the layer of air space between it and Transition Altitude is called the **Transition Layer**. The various altimeter settings are illustrated in Fig. 53.

STANDARD
SETTING
"flight level"

TRANSITION   ALTITUDE

Fig. 53   Altimeter settings and their meanings.

Horizontal Pressure Changes and the Altimeter

Winds flow in a clockwise direction around a high-pressure system and anticlockwise around a low (pages 101 and 92) and on this premise a Dutchman named Buys Ballot has established **Buys Ballot's Law** which in effect states:

'In the northern hemisphere if you stand with your back to the wind the area of low pressure is on your left.'

The significance of this statement and its effect upon the behaviour of the altimeter may best be understood with reference to Fig. 54.

3 Meteorology

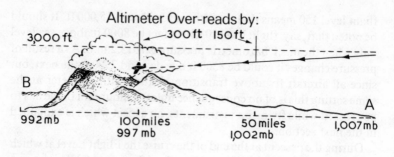

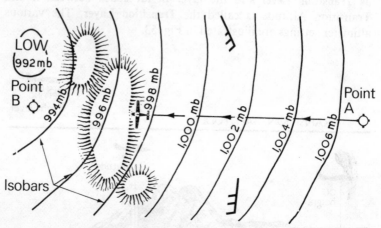

Fig. 54   Flying towards low pressure and the risks of an over-reading altimeter.

This illustrates an aircraft on a cross-country flight which passes over a 2740 ft ridge of hills. The pilot has decided to fly at an altitude of 3000 ft, only allowing 260 ft terrain clearance over the highest point. Had he studied the weather chart before the flight the pilot would have seen that his track led into a region of lower barometric pressure than that at the point of departure. The pilot maintains a constant indicated altitude of 3000 ft but owing to the decreasing pressure the aircraft assumes a gradual descent.

A change in pressure of 1 m.b. will alter the reading on the altimeter by 30 ft. In this case after flying 50 miles the m.s.l. pressure has dropped from 1007 m.b. at the airfield of departure (Point *A*) to 1002 m.b., i.e. a pressure decrease of 5 m.b. or 5 × 30 = 150 ft **Over-read**. After the next 50 miles there is a further pressure

124

reduction to 997 m.b. or 10 m.b. below the original altimeter setting of 10 m.b. or 10 × 30 = 300 ft **Over-read**, so that although a steady 3000 ft reading is maintained on the altimeter the aircraft has in fact descended to 2700 ft a.m.s.l. and is therefore unable to clear the range of hills which on this occasion happens to be in cloud. As a guiding principle it is worth remembering that according to Buys Ballot's Law an aircraft experiencing starboard drift is flying towards an area of lower pressure and the altimeter will over-read (Fig. 55).

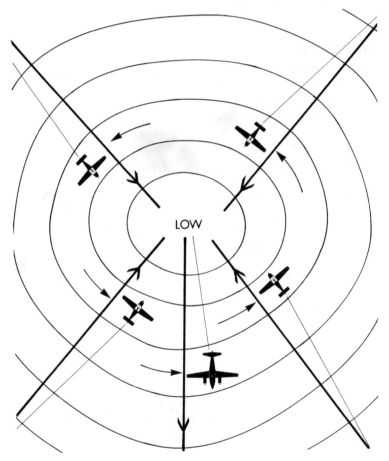

Fig. 55   Buys Ballot's Law.
All single-engine aircraft are flying towards an area of low pressure and have starboard drift. The twin, which is experiencing port drift, is heading away from low pressure towards high pressure.

Conversely when there is port drift the aircraft is flying into higher pressure and the altimeter will under-read so that starboard drift may reduce terrain clearance to the point of danger while port drift will increase the safety margin between aircraft and high ground. These rules are reversed in the southern hemisphere because wind circulation around pressure systems is in the opposite direction to that north of the equator.

Questions

1 **Environmental Lapse Rate is temperature change with height. It changes at a rate of:**
   (a) 2°F per 1000 ft,
   (b) Is dependent on the humidity of the air mass,
   (c) 2°C per 1000 ft.

2 **What is the meaning of the term dew point?**
   (a) The humitidy in a frontal air mass,
   (b) The temperature at which further cooling will cause condensation,
   (c) The point at which the air ceases to rise on reaching a similar environmental temperature.

3 **When air rises there is a drop in temperature due to expansion which together with the lapse rate gives a total lapse rate which is known as:**
   (a) Adiabatic lapse rate,
   (b) Dry adiabatic lapse rate,
   (c) Saturated adiabatic lapse rate.

4 **The dry adiabatic lapse rate is:**
   (a) A temperature drop of 2°C per 1000 ft,
   (b) A temperature drop of 3°C per 1000 ft,
   (c) A temperature drop of 2°F per 1000 ft.

5 **Cirrostratus is:**
   (a) A high-level thin, veil-like cloud,
   (b) Low-level thin cloud cover in which icing is often present,
   (c) A high-level shapeless wispy cloud formation.

6 Thunder and severe turbulence is associated with:
(a) Cumulonimbus clouds,
(b) Nimbostratus clouds,
(c) Altocumulus clouds.

7 A wind is said to veer when there is:
(a) A clockwise change in direction,
(b) A reversal in wind direction,
(c) An anti-clockwise change in direction.

8 The earth's surface has this effect on lower winds:
(a) It produces smoother flying conditions,
(b) It causes a reduction in wind speed, possibly with turbulence,
(c) It affects the wind speed but not its direction.

9 Radiation fog may form at night under the following conditions
(a) Clear sky with a normal adiabatic lapse rate and a moist air mass,
(b) Clear sky, a gentle wind and a moist air mass,
(c) Moist air mass, reasonable cloud cover and a dew point that is easily reached.

10 At ground level the circulation of wind around a depression in the Northern Hemisphere is:
(a) Clockwise blowing into the centre at approximately 30° to the isobars,
(b) Anti-clockwise blowing parallel to the isobars,
(c) Anti-clockwise blowing into the centre at approximately 30° to the isobars.

11 Airframe icing will only occur:
(a) In cloud below freezing level,
(b) In cloud at any ambient temperature,
(c) In any weather conditions below freezing temperature.

12 Airframe icing will affect:
(a) Effectiveness of controls, stalling speed and airspeed,
(b) Effectiveness of controls, airspeed but not stalling speed,
(c) Will decrease the stalling speed and increase the weight.

13 When flying in cloud at an outside air temperature of + 6°C:
   (a) Airframe icing may be expected,
   (b) Carburettor icing may occur,
   (c) There is no risk of airframe or carburettor icing.

14 An aircraft has been left out overnight. It is covered with an icy film which is known as:
   (a) Hoar frost,
   (b) Glazed ice,
   (c) Rime ice.

15 Flight within a cumulonimbus cloud may entail:
   (a) Moderate icing with some precipitation,
   (b) Severe turbulence, icing, lightning and hail,
   (c) Little turbulence, no precipitation but a risk of lightning strike.

16 There is an easterly wind blowing across a range of hills lying North and South. A pilot flying over is likely to notice down draughts on:
   (a) The windward side,
   (b) The leeward side,
   (c) On both the windward and leeward sides.

17 The standard barometric pressure is:
   (a) 1002.3 mb,
   (b) 1003.2 mb
   (c) 1013.2 mb.

18 After landing an altimeter set on the QFE will always read:
   (a) Height above sea-level,
   (b) Zero,
   (c) Airfield elevation above msl.

19 The QNH is related to:
   (a) Altitude,
   (b) Height,
   (c) Flight level.

20 When an aircraft reaches transition altitude the pilot should change his altimeter setting to:
   (a) QFE,
   (b) QNH,
   (c) The standard setting.

21 Buys Ballot's Law states that in the Northern Hemisphere if you stand with your back to the wind the area of lower pressure is:
   (a) On the right,
   (b) On the left,
   (c) Ahead.

22 A change in pressure of 1 mb will alter the reading of an altimeter by:
   (a) 30 ft,
   (b) 60 ft,
   (c) 90 ft.

23 When a large anti-cyclone persists during summer in the British Isles the weather during the day is likely to be:
   (a) Fine,
   (b) Low cloud with rain,
   (c) Thunderstorms.

24 When large cumulus clouds develop which type of precipitation may be expected?
   (a) Drizzle,
   (b) Heavy showers,
   (c) Hail.

25 Poor weather conditions exist but a cold front is forecast to move across the area. When the front has gone through the weather will:
   (a) Improve,
   (b) Deteriorate,
   (c) Remain much the same that day.

26 An aircraft is experiencing starboard drift while flying in the Northern Hemisphere. This indicates that:
   (a) The wind is increasing,
   (b) The aircraft is flying towards an area of low pressure,
   (c) The aircraft is flying towards an area of high pressure.

27 Dull weather with continuous drizzle prevails but a warm front is expected to pass through and afterwards flying conditions will:
   (a) Be much improved,
   (b) Include some drizzle and poor visibility,
   (c) Include a risk of thunderstorms.

3 Meteorology

28 When flying in turbulence it is important to:
(a) Increase the airspeed,
(b) Decrease the airspeed,
(c) Use 10° of flap to lower the stalling speed.

29 The lines joining positions of equal pressure on a synoptic chart are called:
(a) Isobars,
(b) Contour lines,
(c) Millibars

30 An occluded front is:
(a) A poorly developed inactive cold front,
(b) A conglomerate of a cold and warm front,
(c) A poorly developed warm front.

31 The wind was 270/20 kt, which of the following has veered?
(a) 240/20 kt,
(b) 300/20 kt,
(c) 270/10 kt.

32 Advection fog is caused by:
(a) Industrial smoke mixing with moist air,
(b) A moist air mass being lifted over high ground,
(c) A moist air mass drifting over a colder surface.

33 The Station Circle is found:
(a) On a Synoptic Chart,
(b) In a meteorological forecast,
(c) On a route forecast.

34 What are super-cooled water droplets?
(a) Water droplets that form into snow while falling through cloud,
(b) Water droplets that remain in the liquid state at below freezing temperature and which freeze on impact with a surface,
(c) Water droplets that cause rime ice.

35 A series of closely-spaced isobars on a weather map indicates:
(a) High winds,
(b) Low winds,
(c) Possibility of a temperature inversion.

Chapter 4
Principles of Flight

Since the beginning of time, Man has cast envious eyes on the birds of the air, but it is only within living memory and with the emergence of the petrol engine that powered flight has been possible.

For many years before the Wright Brothers made the first powered flight in 1903, a considerable fund of aerodynamic data was being amassed in various countries. There were the Otto Lilienthal experiments with gliders in Germany and similar research by Percy Pilcher in Britain. Even earlier in fact between 1799 and 1810, the foundations of modern aerodynamics were being laid by Sir George Cayley in England. To many about to take up flying, the fascinating principles which make flight possible are as great a source of mystery as they were to the pioneers of a century ago.

These notes are written with a view to explaining, without entering into mathematics, how an aeroplane flies. Obviously in the course of such an explanation simplicity must not be achieved at the expense of accuracy.

'As Light as Air'

The phrase 'as light as air' is a common one, yet air has weight. Imagine a cubic foot of air. If it were possible to place it on a pair of accurate balances in a vacuum it would weigh in the region of $1\frac{1}{3}$ oz. Indeed the air in an average-size room would weigh 130 lb and a large hall may contain several hundredweight. Were it not for this fact an aeroplane would be incapable of flight.

A small piece of metal weighing $1\frac{1}{3}$ oz can be supported without effort on the palm of a hand; fire this same piece of metal from a gun and it will penetrate three or four feet of timber. What has happened to give it such power? Its rapid movement has given it **Momentum** and the ability to exert tremendous force. When moved at speed the $1\frac{1}{3}$ oz cubic foot of air will also gain momentum and it too will generate a force on meeting an object, the power and direction of the force being dependent upon the speed of airflow and the shape of the object.

This force, when coerced in the right directions by the aircraft designer, makes it possible for an aeroplane to fly, be it a light plane or a 300 ton transport aircraft.

How an Aircraft Flies

Before the principles of flight are described in greater detail it may help the student to have an outline understanding of how an aircraft flies.

Lift

The unexpected happens to air (or for that matter any fluid such as water) when it is made to flow around objects of a certain shape.

An interesting experiment with two sheets of stiff notepaper bent across the middle will illustrate this. By holding the two pieces as shown in Fig. 56 with their creases approximately one inch apart they can be made to close together by blowing hard between them.

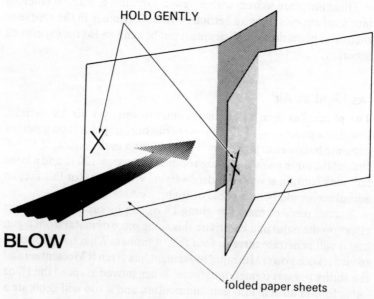

HOLD GENTLY

BLOW

folded paper sheets

Fig. 56   An experiment with two sheets of folded paper. Blow between them and they close together.

In effect the impossible would seem to have happened. The explanation is as follows: the air is forced through a passage which progressively becomes narrower. In order to pass through the restriction it must increase speed (Fig. 57). A law of nature demands that when a moving fluid is forced through a restriction in this way, the increase in speed is accompanied by a decrease in pressure. It is this drop in pressure which causes the sheets of paper to close together.

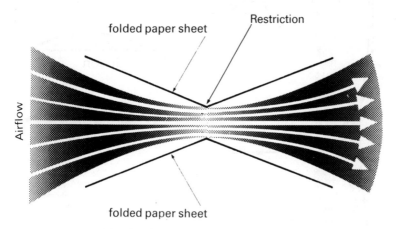

Fig. 57   The folded paper in the previous illustration seen from above. To pass through the restriction the airflow must increase speed, a decrease in pressure follows and the paper does the unexpected; the sheets close together.

A similar experiment can be made with a large tablespoon. If this is held downwards and brought into contact with a running tap, the water will pull the spoon into the jet although one would expect the spoon to be pushed away (Fig. 58). Here again the fluid (in this case water) is made to flow around the contours of the spoon, causing a decrease in pressure.

A shape almost identical to the back of a spoon is incorporated in the wing of an aeroplane and a section cut through its width would reveal a remarkable similarity.

The shape which is shown in Fig. 59 is called an **Airfoil Section** and it is the basis upon which flight depends.

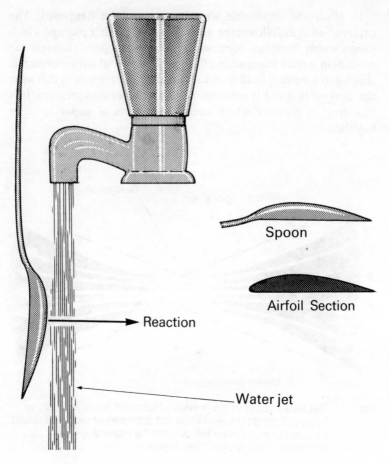

Spoon

Airfoil Section

Reaction

Water jet

Fig. 58   A similar experiment to that shown in Figs. 56 and 57, this time
with a single surface – the back of a spoon. Instead of
pushing it away the water pulls in the spoon. Compare its
profile with the typical Airfoil section shown on the right.

Behaviour of an Airfoil Section

By introducing smoke into a jet of air it is possible to watch the
behaviour of the streams passing above and below a model wing
section placed within a wind tunnel (Fig. 60).

The previous paragraph explained that a drop in pressure occurs
on the top surface, but if the **Leading Edge** of the airfoil is raised at a
slight angle to the airflow, because of its momentum, pressure will

rise when the air makes contact with the undersurface of the wing (Fig. 61).

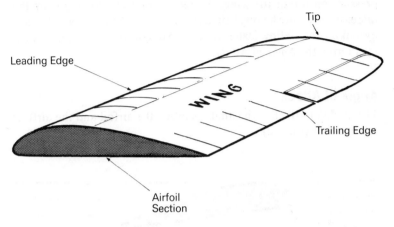

Fig. 59   A wing showing its Airfoil Section which is the essential building block of flight.

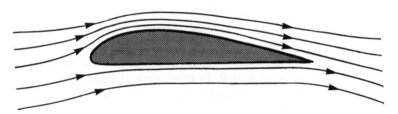

Fig. 60   Flow of air around an airfoil.

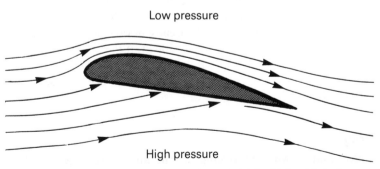

Fig. 61   Build-up of high pressure under the airfoil which adds to the force caused by the decrease in pressure on top.

135

The net result is that the airfoil section will generate a lifting force. Approximately two-thirds of this is contributed by the decrease in pressure on top of the wing, the remaining third coming from the increased pressure below. **Lift**, as the force is called, can be increased by making the air flow faster. It can also be controlled in another way, by altering the **Angle of Attack**.

Angle of Attack

The angle of attack is the angle between the airfoil and the airflow relative to it (Fig. 62).

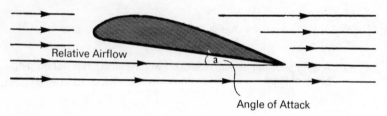

Relative Airflow

Angle of Attack

Fig. 62   The Angle of Attack.

It should be emphasized that the angle is between the *relative* airflow and the airfoil so that the same angle of attack could occur under differing flight conditions such as climb, straight and level flight, or gliding (Fig. 63).

Lift is generated from most parts of the airfoil and if it is measured and represented pictorially the distribution of force would be as shown in Fig. 64. It is more convenient to depict these individual forces in one line drawn as the point from which the total effect occurs. This point is known as the **Centre of Pressure** (Fig. 65).

Assuming a steady airflow of say 100 kt, if the angle of attack is increased then the amount of lift will increase. At the same time the centre of pressure moves forward – the significance of this will be explained later.

Drag

Unfortunately while producing lift the airfoil also creates a less desirable force known as **Drag**. Clearly any body in a moving fluid such as air must cause resistance and while the lift acts at right angles

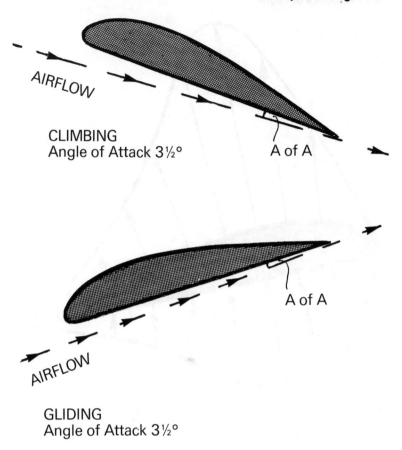

CLIMBING
Angle of Attack 3½°

A of A

A of A

GLIDING
Angle of Attack 3½°

Fig. 63   The angle of attack is measured from the Relative Airflow,
consequently the same angle may occur in level, climbing and
gliding flight.

to the relative airflow, drag will of course be parallel to it (Fig. 66).
Because it is a by-product of lift generation this particular form of air
resistance is known as **Induced Drag**. However, all parts of the aircraft
that are exposed to the airflow cause drag and the subject will be
explained later in the chapter.

The functions of lift, drag and the centre of pressure should now be
considered under various angles of attack. In the diagrams the
direction of force is shown by the arrows and their length indicates
the amount of force. It can be seen in Fig. 67 that, as the angle of
attack increases, the lift force becomes more powerful and the centre

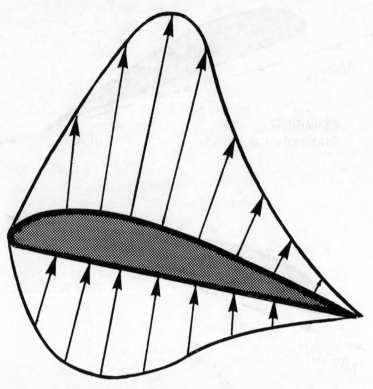

Fig. 64   Pressure distribution envelope around an airfoil.

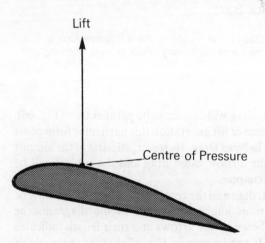

Fig. 65   Lift, represented by a single line as the sum of all forces shown
          in Fig. 64, acts from the Centre of Pressure.

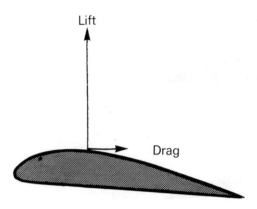

Fig. 66 Drag acts at 90° to Lift.

of pressure moves forward. At the same time drag increases slowly at first up to approximately 4° but more rapidly as the angle increases. The greatest amount of lift for the least amount of drag is known as the **Best Lift/Drag Ratio** and it usually occurs at $3\frac{1}{2}$–4°.

A point is reached when further increases of angle will produce no more lift. Taking the extreme case, if the wing were to be forced through the air at 90° angle of attack there would be no lift at all, only a considerable amount of drag. At some point between 0° and 90° a marked deterioration occurs. This is known as the **Stalling Angle** and it is dealt with more fully in Exercise 10A, Stalling, *Flight Briefing for Pilots, Vol. 1.*

Factors affecting Lift

These are the factors which affect the amount of lift a wing can generate –

1 *Angle of Attack.* At any particular airspeed an increase in angle up to the stalling angle gives more lift.

2 *Airspeed.* For a particular angle of attack the faster the airspeed the more lift for a particular wing.

3 *Airfoil section.* There are many variations of these shapes each designed for a purpose (Fig. 68). By and large, the deep, highly cambered airfoils give the most lift for a given speed while thinner sections are used when a high cruising speed is required of the design.

4 *Wing area.* The larger the area of wing of a given airfoil section the more weight it will support at any particular speed.

4 Principles of Flight

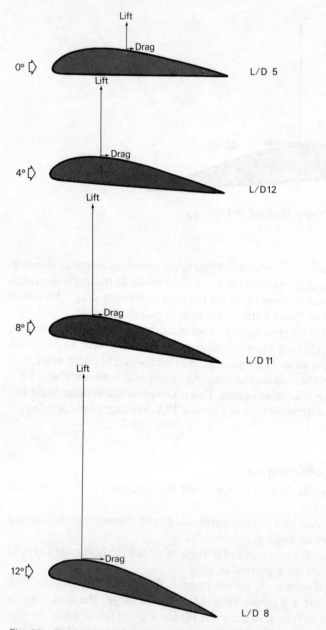

Fig. 67   The effect of angle of attack on lift and drag. The forces, as
depicted by the length of their arrows, are drawn to scale.
Figures on the right refer to the Lift/Drag Ratio of a typical
aircraft.

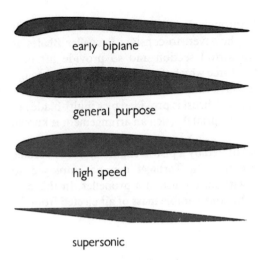

early biplane

general purpose

high speed

supersonic

Fig. 68   Some typical airfoil sections.

5 *Air density*. The weight of air varies with height and temperature, being little more than half the sea level density at 20,000 ft. The more dense the air the more lift a particular wing wil give for a particular airspeed. It is rather like the difference between swimming in fresh and sea water; sea water being denser than fresh water, offers more support.

Creating the Airflow

For the purpose of explanation the airfoil has been assumed to be stationary with an airflow moving over it. In practice this is not the case. Bearing in mind that it is the *relative airflow* which provides lift it will be realized that, by moving the wing through the air at speed, lift and drag will occur in exactly the same way. In a helicopter this is arranged by having two or more wings rotate around a central point with a linkage to control the angle of attack. In a fixed-wing aircraft the relative airflow is generated by pulling (or pushing) the entire machine through the air.

Forward motion achieved by driving the wheels like a car would terminate as soon as the aircraft lifted off the ground. Instead the engine rotates an **Airscrew** (popularly known as the Propeller). Explained in simple terms airscrew describes its function well, since it does in fact move itself through the air like a large wood screw. It would be more accurate, however, to consider **Propeller Blades** as wings since they are of airfoil section and so provide lift in a horizontal plane. It is this force which thrusts the aeroplane through the air: indeed, the force which causes the aeroplane to move forward is called **Thrust**. In a few cases thrust is provided by a multi-blade fan rotating within a duct. Appropriately such an arrangement is known as a **Ducted Fan**.

The propeller may be rotated by a piston engine or a **Gas Turbine (Turbo-propeller)**. Alternatively a **Turbojet** can be employed to produce thrust directly without the use of a propeller. In this case thrust is the reaction to the considerable mass of air ejected from the rear of the engine.

The Aeroplane

The parts of the aircraft explained in the preceding pages may now be assembled.

First, there is the wing of airfoil section. In order to generate lift, a means of propelling it through the air must be attached, in this example a petrol engine driving an airscrew (Fig. 69).

This strange assembly is incapable of controllable flight because it lacks a characteristic which is of prime importance to any vehicle whether motor-car, boat or aeroplane – **Stability**. Taking a simple case the term can be applied to a handcart. With the greatest of difficulty it would be possible to balance the vehicle illustrated in Fig. 70, but the slightest tilt would become more and more

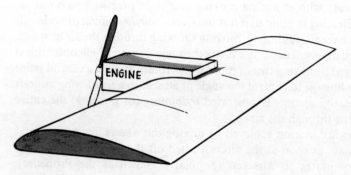

Fig. 69   The wing with a means of populsion added.

Fig. 70   Like the 'aircraft' shown in Fig. 69 this is an unstable vehicle.

pronounced, with disastrous results. The airfoil section is equally unstable. Referring back to the behaviour of the centre of pressure, it will be realized that for this strange craft to remain in balance its centre of gravity must coincide with the centre of pressure.

Imagine the aeroplane is flying with its lift and weight in balance as shown in Fig. 71 when a disturbance in the air lifts the front of the machine slightly, thus increasing the angle of attack. The immediate effect on the centre of pressure has already been mentioned: it will move forward.

If the two lines of force marked lift and weight in Fig. 72 are imagined to be lengths of string pulling in the direction of the arrows, it can be seen that by moving forward, the centre of pressure has added to the nose-up disturbance.

The two forces now produce a 'nose up' motion which still further increases the angle of attack. This in turn causes the centre of pressure to move forward again and the machine will commence a series of uncontrollable rotations.

The remedy is simple. In the case of the handcart, shafts are incorporated as in Fig. 73. Part of the function of the 'driver' is to stabilize the handcart through these shafts. A remarkably similar arrangement is used by the aircraft designer. To the wing is attached a **Fuselage:** this not only acts as the shaft of the handcart but also houses the engine, pilot and passengers and in many cases other items such as fuel and freight. In place of the man at the end of the shaft is a

143

4 Principles of Flight

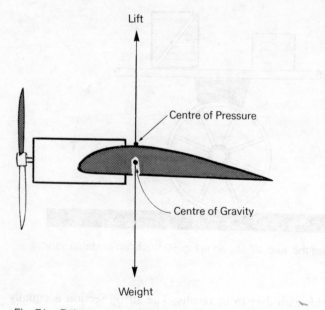

Fig. 71   Balanced on a knife edge.
This arrangement is as unstable as the cart shown in Fig. 70.

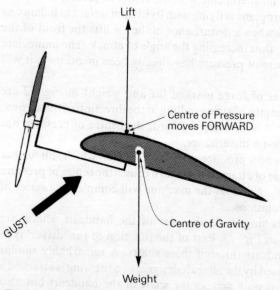

Fig. 72   Reason for instability. A nose-up tilt caused by a gust results in
forward movement of the centre of pressure and a further
nose-up tendency.

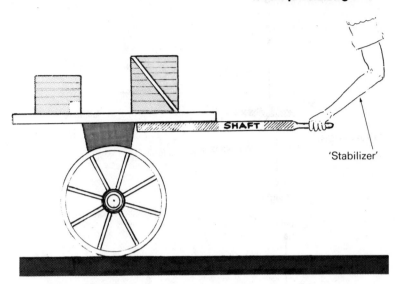

Fig. 73   Simple cure for an unstable vehicle.

small surface called the **Tailplane**. Its function is to stabilize the behaviour of the **Mainplane** (wing). In America the tailplane is referred to as the Stabilizer.

It is the tailplane which gives the aircraft **Longitudinal Stability**. Usually the tailplane is of a symmetrical airfoil section giving no lift when at 0° angle of attack. Should a disturbance cause the nose to rise, the tailplane assumes a positive angle and produces lift which, acting through the leverage of the fuselage, will bring the mainplane back to its original position. In the event of a gust causing the nose of the aircraft to drop, the tailplane will assume a negative angle of attack and produce a corrective force in a downward direction. Reference to Fig. 74 will make the function of the tailplane clear.

The fuselage, wing, tailplane and method of propulsion so far described would be incapable of safe flight because, although the needs of longitudinal stability have been satisfied, an aircraft must also be stable in direction (**Directional Stability**) and it should not continually roll from one wingtip to the other. To prevent this an aircraft is designed to have **Lateral Stability**.

For an aircraft to be of practical use it must be controllable; the pilot will need to turn, raise or lower the nose, level or bank the wings and so forth. An aircraft is therefore provided with **Flying Controls**.

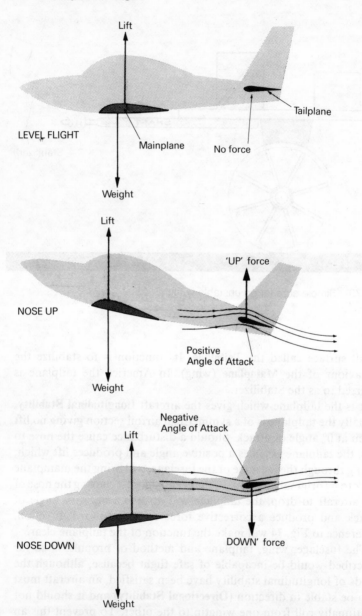

Fig. 74   Aerodynamic equivalent to the shafts of a handcart and its 'stabilizer' are the fuselage and the tailplane. In level flight there is no force from the tailplane which usually has a symmetrical airfoil section. A nose-up or nose-down disturbance causes the tailplane to generate a correcting force.

These, and stability as mentioned in the preceding paragraph, are described in the relevant sections that now follow.

Stability

Unlike most other vehicles an aeroplane has freedom of movement in three planes:

It can **Pitch**, nose up, nose down.
It can **Roll**, left wing down, right wing down.
It can **Yaw**, nose to the left, nose to the right.

These three movements each occur around a separate axis which passes through the aircraft's centre of gravity (Fig. 75). When it departs from **Straight and Level Flight** an aircraft:

pitches around its **Lateral Axis**,
rolls around its **Longitudinal Axis**,
yaws around its **Vertical Axis**.

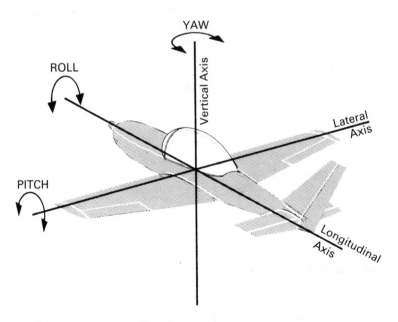

Fig. 75   The three planes of movement of an aircraft around the Lateral Axis, Longitudinal Axis and Vertical Axis.

The stability of an aircraft is very much a design problem. Nevertheless a well-informed pilot should be conversant with the basic principles of stability.

The object of Stability

When, after a disturbance, an aeroplane returns to its original attitude without any corrective action on the part of the pilot, it is said to possess stability. For example, an aeroplane in straight and level flight may be deflected into a nose-down attitude during turbulent flying conditions. Were the nose to drop still further in a dive of increasing steepness the aeroplane would be **Unstable**. Both stability and instability may be related to movement around all three axes and, while up to a point stability is a desirable characteristic in an aeroplane, instability can be dangerous and much research has gone into its elimination by good aerodynamic design.

Stability characteristics

The tendency for an aeroplane to return towards its original trimmed condition of flight after it has been displaced is called **Static Stability**. This in itself is insufficient since in making the correction the aircraft follows an undulating path and behaves rather like a motor vehicle without shock absorbers.

The damping of these oscillations necessitates **Dynamic Stability** and the corrective behaviour of an aeroplane which is statically stable in the pitching plane is compared with a dynamically stable aircraft in Fig. 76.

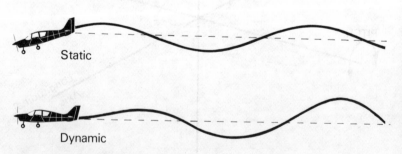

Static

Dynamic

Fig. 76   Difference between Static and Dynamic Stability, in this case illustrated in the pitching plane although it relates to stability in roll and yaw.

In addition to these main characteristics, stability may vary in effectiveness from one type to another. A large transport aircraft will possess a high degree of stability whereas in a training aircraft this may be partially sacrificed in the interest of manoeuvrability and control.

Methods of attaining Stability

Corrections in the pitching plane or longitudinal stability were earlier explained. A disturbance in the yawing plane is corrected with directional stability and corrections in the rolling plane require lateral stability. Some of the design features incorporated in attaining stability are explained under their relevant headings.

Longitudinal Stability

It was previously explained that an airfoil is unstable because its centre of pressure moves forward as the angle of attack increases and vice versa. To overcome the out of balance **Couple** which occurs when lift and weight become out of alignment, a tailplane is fitted. The tailplane must correct the disturbance which was the prime cause of the change in attitude and the unstable movement of the centre of pressure.

After a disturbance the aircraft will, because of inertia, continue momentarily in a nose-up or nose-down attitude without changing its flight path. During the temporary change in attitude the tailplane assumes either a positive or negative angle of attack, so effecting a correction.

The degree of longitudinal stability is influenced by the following factors:

1. Movement of the Centre of Pressure

Modern airfoil sections are designed so that the centre of pressure remains within fairly close limits so reducing the magnitude of any nose up or down couple and improving longitudinal stability.

2. Area of Tailplane and length of Fuselage

A large tailplane will exert more force than one of smaller area so that corrective action will be more positive. A long fuselage provides the tailplane with extra leverage, therefore, tailplane size being equal, a long fuselage will assist longitudinal stability.

149

3. Position of the Centre of Gravity

This is an important factor since it is to some extent under the control of the pilot. In principle when the centre of gravity is well forward, longitudinal stability is at its best, although a point can be reached when elevator control becomes heavy and difficult, particularly at low speeds. As the centre of gravity is brought aft, longitudinal control improves at the expense of stability until a position is reached when instability occurs.

Centre of Gravity Limits are quoted in the Owner's/Flight/Operating Manual for all aircraft and it is the pilot's responsibility to ensure that these are not exceeded when loading the aircraft. Various methods are in use to determine the position of the centre of gravity and while little can go wrong with the loading of a small 2/4 seat aeroplane, the correct disposition of the load is of prime importance with larger types. Methods of determining **Weight and Balance** will be shown in Section 6 of the Owner's/Flight/Operating Manual for the aircraft. The subject is explained in Chapter 8 of this book.

Lateral Stability

The close relationship which exists between lateral and directional control is clearly demonstrated in the two exercises described in Volume 1 of this series 'Further Effects of Aileron and Further Effects of Rudder'. Likewise there exists a close link between lateral and directional stability, although for the purpose of explanation these are considered separately.

Lateral stability is attained by using one or more of the following methods:
1. High Wing
2. Dihedral Angle
3. High Keel Surface
4. Sweepback

1. High Wing

By situating the wing on top of the fuselage, well above the centre of gravity, there will exist a natural tendency for the aircraft to remain level rather like a plumb line. After lateral level has been disturbed the centre of gravity swings into alignment with the centre of lift so levelling the wings. This action is called **Pendulous Stability** (Fig. 77).

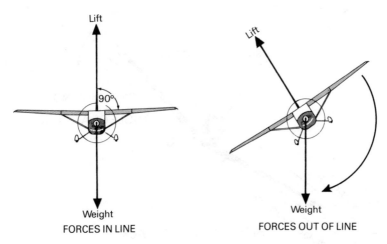

Fig. 77   Pendulous Stability. A high-wing monoplane derives lateral
stability through pendulum action. Since Lift acts at 90 degrees
to the wing when lateral level is disturbed (right-hand picture)
Lift and Weight are out of alignment and a righting couple is
caused.

2. Dihedral Angle

When the wing tips are raised the angle so formed between the wings
and the horizontal is called dihedral (Fig. 78). When as a result of
other design factors too much lateral stability exists to the detriment
of lateral control, the wing tips are lowered below the horizontal to
form an **Anhedral Angle** but its use in light aircraft is very unusual.
For its corrective action, dihedral angle is dependent upon the
sideslip which occurs when a wing goes down for any reason not
associated with yaw. The sideslip causes the relative airflow to change
direction and, because of dihedral, meet each wing at a different angle
of attack. The illustration shows that lateral stability is achieved by:

(*a*) the larger angle of attack on the lower wing during the sideslip;

(*b*) loss of lift on the uppermost wing due to the partial shielding
effect of the fuselage.

3. High Keel Surface

The term refers to all side area which is above the centre of gravity
and once again sideslip provides the corrective airflow. When more
keel area exists above the centre of gravity than below, there will be a

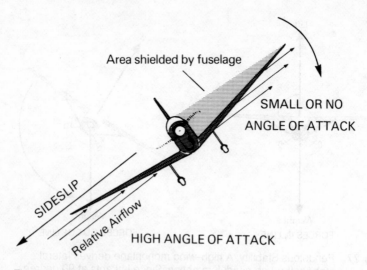

Area shielded by fuselage

SMALL OR NO
ANGLE OF ATTACK

SIDESLIP

Relative Airflow

HIGH ANGLE OF ATTACK

Fig. 78   Dihedral Angle provides lateral stability by presenting the wings
at different angles of attack during a wing-down/sideslip
situation.

levelling tendency during a sideslip (Fig. 79). Use of high fin as a
means of contributing to lateral stability can have undesirable side
effects. These are outlined under 'Stability Interaction'.

4. Sweepback

Yet another method of attaining lateral stability is **Sweepback**; seen in
plan view the wings angle back from where they join the fuselage
towards the tips.

This method is confined to high speed jet aircraft and it need not
concern the private pilot.

Directional Stability

When the principles governing the behaviour of a weathercock are
understood, directional stability as it applies to the aeroplane will
need little explanation. The requirement for both weather vane and
aeroplane is the same – more keel area behind the turning point than
in front. In the weather vane, this turning point is represented by the
pivot around which the indicator 'veers' or 'backs' according to the
wind, whereas the turning point of an aircraft is its centre of gravity.

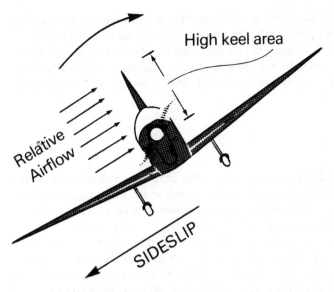

Fig. 79 A high fin can contribute to lateral stability although there may be undesirable side effects.

Although fuselage keel area is greater behind the centre of gravity than in front in most (but not all) aircraft, to provide an acceptable degree of directional stability this is augmented by the addition of a **Fin** (Fig. 80). Like the tailplane and its contribution towards longitudinal stability, the fin effectiveness is influenced by fuselage length, i.e. fin area being equal, a long fuselage behind the centre of gravity enjoys increased leverage and therefore gives better directional stability than one of shorter length.

Stability Interaction

Because of the relationship already mentioned between directional and lateral control and stability, certain complications must be avoided by the designer in his quest for stability. The function of the sideslip in lateral stability may cause a strong turning motion towards the dropped wing, particularly when a large fin is used to promote directional stability. The turn causes the raised wing to gain lift because being on the outside of the turn it is flying on a slightly greater radius and therefore has slightly greater speed. When this is sufficient to overpower the effects of dihedral in levelling the wings, a

153

spiral dive will develop which becomes progressively steeper. Such a condition is called **Spiral Instability** and its design treatment is usually confined to reducing the fin area until its effects are at an acceptable level. It should be explained that most modern aircraft have a slight tendency towards spiral instability and, in fact, small amounts assist in turning.

Drag

In the section at the beginning of this chapter, which explained in the broadest terms how an aircraft was able to fly, it was mentioned that while generating lift the wing also produced drag. In the main, this is **Induced Drag**, a by-product of the lift-generating process.

Drag is also created by other parts of the **Airframe**. There are the fuselage, tail surfaces and indeed the wings themselves to be propelled through the air and, in much the same way as a boat moving through the water, resistance will be generated. Whereas induced drag is the direct result of the wing creating lift, **Profile Drag**, air resistance

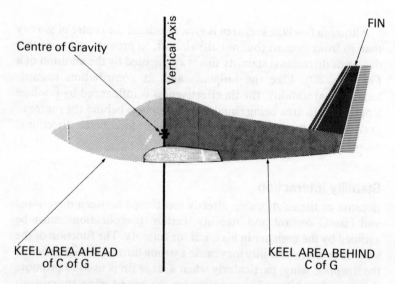

Fig. 80   Keel area behind the centre of gravity, being greater than that before, the aircraft has a natural tendency to 'weathercock' into the airflow. This, coupled with the Fin, provides an aircraft with directional stability.

caused by the airframe, is the enemy of flight in so far as it hinders movement through the air. Unless the designer has been successful in minimizing it, power is dissipated in overcoming profile drag instead of endowing the aircraft with a high cruising speed and a good rate of climb.

1. Profile Drag

Whenever a body is moved through air (or water) resistance will result from two sources:

(*a*) **Form drag** caused by the body as it disturbs the air through which it passes, and

(*b*) **Skin friction** which is incurred as the air passes over and under the surface of the body.

Both of these aspects of profile drag are influenced by good design and manufacture and their reduction gives increased performance.

For practical purposes profile drag increases as the square of the indicated airspeed, e.g. twice the speed, four times the drag; three times the speed, nine times the drag, etc., and the following figures clearly illustrate the detrimental effect of this type of drag particularly as the airspeed increases.

Profile Drag for Aeroplane X

Indicated Airspeed	Profile Drag
100 kt	200 lb
200 kt	800 lb
300 kt	1800 lb
400 kt	3200 lb
500 kt	5000 lb
600 kt	7200 lb

Each pound of profile drag must be paid for in terms of equal thrust and during the design stage every effort is made to reduce both form drag and skin friction. For example a 10 per cent reduction would mean a comparable saving in power at each airspeed. The designer could either decide upon a smaller engine with an attendant reduction in weight, or he could exploit the saving in drag and have a higher all-round performance with the engine originally selected. Indirectly a saving in drag represents fuel economy and this in turn influences the aircraft's range so that much time and effort is devoted to the reduction of profile drag. Because of its nature it is sometimes called

4 Principles of Flight

Parasite Drag and some of the methods used to combat its two contributing factors are listed below.

Form Drag

Other things being equal, shape is all important since by definition this kind of drag results from the disturbance created by a body as it passes through the surrounding air. Minimize the disturbance by streamlining and form drag will be reduced. The effect of streamlining is shown in Fig. 81.

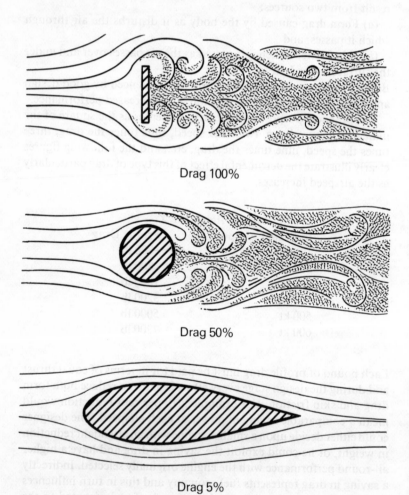

Drag 100%

Drag 50%

Drag 5%

Fig. 81   Reducing drag by streamlining.

Most disturbance is caused when a flat plate is forced through the air in the manner shown in the diagram, but by correct streamlining its form drag can be reduced to one twentieth of the original value.

While a gradual change in the direction of airflow is conducive to low form drag, this entails long thin shapes of considerable surface area and in the more detailed explanation which follows it will be seen that area causes skin friction. Because of these conflicting requirements the ratio between the length and thickness of a streamlined body is important. This aspect of a streamlined body is called **Fineness Ratio** (Fig. 82).

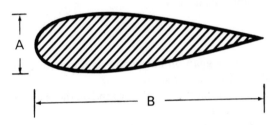

Fig. 82   Fineness Ratio. It is expressed as a number found by dividing measurement 'B' by measurement 'A'.

When dealing with speeds up to 300 kt or so a fineness ratio of 4 is ideal, higher speeds requiring a higher fineness ratio.

While two components of an aircraft may in themselves be of low form drag, when they are joined the airflow over one part may react with the other to produce **Interference Drag**. This is really another type of form drag and a typical example would be the interference drag caused at the point where the wing joins the fuselage. By using **Fillets** one surface is made to flow into the other, thus reducing interference drag (Fig. 83).

Skin Friction

There are many examples of skin friction in everyday occurrence and their mention will contribute much towards an understanding of this form of resistance.

1. The resistance felt whilst shaving, which is reduced by the introduction of some form of lather.

2. The relative ease of operation when anything mechanical is lubricated.

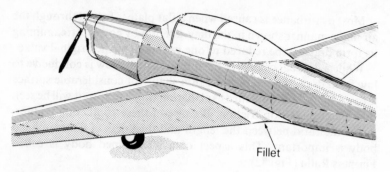

Fillet

Fig. 83   Interference drag between wing and fuselage is often minimised by incorporating a Fillet.

3. The high speed of a toboggan on hard snow as opposed to its immobility on a road free from ice or snow.

4. Wind gradient or the slowing effect of the ground on those layers of moving air in contact with it.

All these examples are self-explanatory and to a large extent find an aerodynamic parallel in skin friction. Because of porosity or other irregularities in the finish of an aircraft the layer of air in immediate surface contact tends to adhere and become stationary relative to the airframe. This **Boundary Layer** extends for up to 5 mm from the surface, its thickness being determined by the degree of smoothness (or otherwise) of the aircraft's finish. While it is true to say that a well-polished surface reduces skin friction, it can never be completely eliminated and the boundary layer will always flow more slowly than those streams of air more removed from the aircraft. Provided the boundary layer can be maintained in a **Laminar** state, i.e. smooth flow, skin friction will be kept within reasonable limits. A poor surface will cause the boundary layer to break up into countless eddies and the resultant **Turbulent Boundary Layer** will thicken and incur an increase in skin friction. Skin friction is responsible for up to 35 per cent of the total drag on a high-speed aircraft.

Over a considerable number of years experiments have been conducted into the possibility of **Boundary Layer Control**. The thought behind these experiments is that skin friction can be practically eliminated provided the boundary layer is kept moving. The experiments have been largely confined to the wing surfaces and areas have been made porous through which the boundary layer is removed by vacuum pumps. The results are said to promise worthwhile reductions in skin friction and boundary layer research continues.

Reducing Profile Drag

Another and less complex method of controlling skin friction is by limiting the **Wetted Area** or total area of the aircraft in contact with the boundary layer. Similarly form drag can be reduced by keeping the **Frontal Area** of the aircraft as small as possible, e.g the cross-section of the fuselage, the size of the windscreen, the use of a retractable undercarriage, etc. It therefore follows that profile drag (made up from form drag and skin friction) must be reduced by designing the aircraft with the smallest possible frontal and wetted area, the smoothest possible surface, the best streamlined shape and freedom from all unnecessary appendages during flight, such as undercarriage legs, nose or tail wheel.

Unfortunately these requirements often conflict with operational demands such as internal accommodation, pilot's view, directional and other stability and cost of manufacture so that most aircraft represent a compromise.

2. Induced Drag

While all parts of the aircraft, including the wings, are subjected to profile drag, an additional form of resistance is present in flight which, together with profile drag, makes up the total drag of the aircraft. This is induced drag to which reference has already been made at the beginning of the chapter. As previously explained it is inseparable from the wing's function as a lift producer although steps can be taken to reduce its effects.

To understand induced drag it is necessary to refer back to basic principles when it will be remembered that in the process of producing lift, high pressure is caused under the wing and low pressure above. Figure 84 shows that in effect the high-pressure area is separated from the low-pressure area by the wing itself.

When a pneumatic tyre is punctured, air escapes into the atmosphere because it has been confined within the tyre at a higher pressure than the surrounding air. Unequal pressures cannot exist side by side unless they are separated by some means, and, unlike a serviceable tyre, a wing is an imperfect pressure separator. The diagram shows that, because of the difference in pressure, air from below the wings flows around the tips into the low-pressure area. The loss of pressure from below and the rise in pressure which results on

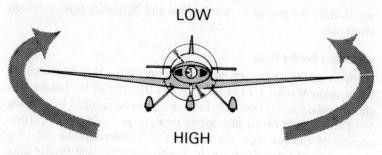

Fig. 84   Wingtip flow of air from high to low pressure.

the top surface has a damaging effect on the lift produced by the area of wing adjacent to the tips, although this complication is additional to the considerations of induced drag. As the air flows around the airfoil from leading to trailing edge, its path is influenced by the circulation (around the wing tips) from high to low pressure and the airstream on the top surface is deflected inwards towards the fuselage while an outward flow towards the wing tips occurs below. The net result amounts to a deviation from parallel flow and Fig. 85 shows that, as the upper and lower air streams meet at the trailing edge, their paths in relation to one another differ slightly setting up a number of rotating eddies, while a particularly large helix is caused at each wing tip. The **Wing-tip Vortices** in conjunction with the smaller **Trailing Edge Vortices** are the cause of induced drag and it follows that they will become more active as the pressure difference increases between upper and lower wing surfaces, i.e. at high angles of attack.

Should the angle of attack be reduced to the point when no lift is produced (usually a minus angle), the vortices will disappear and in consequence there will be no induced drag. Once the angle of attack is increased lift will occur and vortices develop causing induced drag, becoming more pronounced as the angle of attack is increased. As both angle of attack and induced drag increase the airspeed will decrease. In other words induced drag becomes less as speed is increased, whereas the reverse is the case with profile drag.

Reducing Induced Drag

Induced drag may be reduced by efficient design particularly with regard to the shape of the wing. Since the spillage of air around the wing tips, from high pressure to low, causes the vortices responsible for induced drag, it follows that any reduction in circulation around

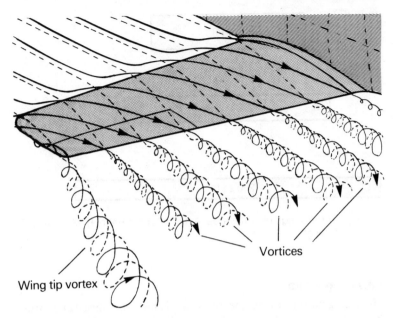

Fig. 85   Induced Drag caused by vortices which are the result of
           spanwise flow (see Fig. 84). Solid lines depict inward flow over
           the top of the wing. Broken lines represent the outward flow
           underneath.

the tips will minimize the effects of induced drag. Assuming that a
wing area of 200 square feet is required for a particular design, this
could be obtained in a number of ways and three alternatives are
illustrated in Fig. 86. The relationship between the span of the wing
and its chord (or average chord when the wing is tapered) is called
Aspect Ratio, wing A being an example of a low-aspect ratio wing,
while wing C is of the high-aspect ratio type associated with high-
efficiency sailplanes. For a given wing area, high-aspect ratio
produces more lift and less induced drag than a low-aspect ratio wing
at the same speed and angle of attack, mainly because of the smaller
wing tips which reduce the air spillage from high pressure to low.
Unfortunately high-aspect ratio wings present structural problems, it
being difficult to construct a wide-span narrow-chord wing of
acceptable stiffness without incurring a severe penalty in weight.

Whereas sailplanes may have an aspect ratio of 30 or more, 8 to 10
is usual for transport aircraft and 5 or 6 for low performance light
aircraft.

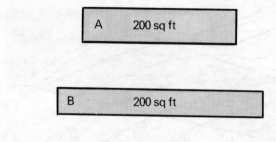

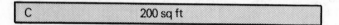

Fig. 86   Three wings of equal area; A. Low Aspect Ratio; B. Medium;
C. High.

Lift/Drag Ratio

Since drag must be overcome by an equal amount of thrust and thrust
has to be provided by engine power it is obviously advantageous to
save drag whenever possible. Some of the design features adopted for
the purpose (good wheel fairings or, better still, retractable
undercarriages, smooth surface, etc.) have already been mentioned.

Sufficient lift to support the aircraft in flight must be provided but
if this can be done with a minimum of drag overall performance is
bound to benefit. Much research has been devoted to designing airfoil
sections that give the most amount of lift for the least drag and Fig. 67
on page 140 illustrates how the lift/drag ratio of a wing changes as its
angle of attack is altered. L/D performance of a wing is the starting
point but it is the L/D ratio of the entire aircraft (i.e. taking into
account the total drag of the airframe) that determines its efficiency.

An average light aircraft would have a L/D Ratio of 8 and a more
efficient one possibly 11. High speed passenger jets can improve on
this figure and some of the better motorgliders, which feature very
'clean' airframes and high aspect ratio wings, produce thirty times
more lift than drag. In practical terms these L/D figures mean that an
aircraft with a ratio of 8 would glide a distance of 8000 ft for every
1000 ft of height loss while the motorglider with its L/D of 30 would
travel 30,000 ft (almost 5 nautical miles) per 1000 ft of descent.

Because, for any particular speed, aircraft with high L/D Ratios
need less power than those of lower ratio, L/D ratio is regarded as an
important measure of efficiency.

Flying Controls

So that an aircraft may be controlled about its three axes the following Primary Controls are provided:

Elevators   for control in the pitching plane
Ailerons   for control in the rolling plane
Rudder   for control in the yawing plane

Use of these three controls and their location on the airframe is described in Exercise 4, Effects of Controls, *Flight Briefing for Pilots, Vol. 1*. In this ground studies manual the subject is described in further detail.

Balanced Controls

In light aircraft and even some quite large designs the rudder, elevators and ailerons are moved by the pilot via a system of cables running over pulleys or rods (explained in Chapter 5, Airframes). The amount of effort required to move these controls against the forces exerted by the airflow increases with airspeed. It is therefore necessary to provide some form of **Aerodynamic Balance** to relieve the pilot of at least part of the effort required while flying the aeroplane.

Balance Tabs

A common method used is the **Balance Tab**, a small surface often inset within the trailing edge of the control and linked so that it moves in the *opposite* direction to the main control when it is moved by the pilot (Fig. 87).

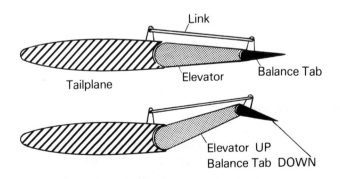

Fig. 87   Balance Tab, a means of reducing control loads.

By moving in the opposite direction the balance tab exerts a force in the same direction as that being applied by the pilot. Balance tabs may be fitted to any of the three primary controls. An alternative method of balancing a control is:

Horn Balance (Fig. 88)

This may be seen on both rudder and elevators on many aircraft. The shaded area in the illustration is added in front of the hinge line of the control surface, so balancing part of the load needed to move the control. Too much area before the hinge line would overbalance the control, so causing the pilot to restrain the control from 'overbalance' when movement is made. On occasion it is desirable to have no assistane during small control movements and a gradual introduction of balance as larger applications of control are made. Such an arrangement is called **Graduated Horn Balance** and it incorporates a shield behind which the balance area is masked from the airflow until larger movements of the control bring it into action.

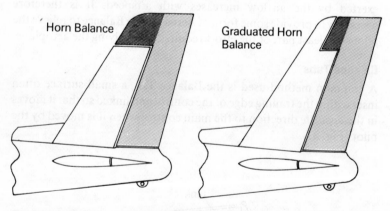

Fig. 88   Horn Balance, another method of reducing control loads, in this example applied to the rudder surface.

All-moving tailplanes

The All-moving Tailplane, which is known as a **Stabilator** in the USA, is, like the swept fin and rudder, a design feature adopted by light aircraft manufacturers from high speed aircraft where they are of more relevance. However, a number of light aircraft have these devices and they are described herewith.

Instead of a fixed tailplane to which are hinged elevators the entire tailplane is made to pivot and change angle (Fig. 89). However this brings with it certain difficulties. Imagine the pilot has moved forward the wheel/stick. The all-moving tailplane would go down and adopt an increased angle of attack, causing its centre of pressure to move forward of the hinge or pivot point. A powerful force would then try and move the control still further. Likewise, when the wheel/stick is moved back and the tailplane adopts a negative angle of attack (to provide a down force and raise the nose of the aircraft), centre of pressure would again move forward, this time below the surface with the force acting downwards. An over-balancing force would likewise try and move the control still further in the direction originally applied by the pilot.

To prevent over-balanced control, a potentially dangerous situation, a small **Anti-balance Tab** is fitted at the trailing edge and so

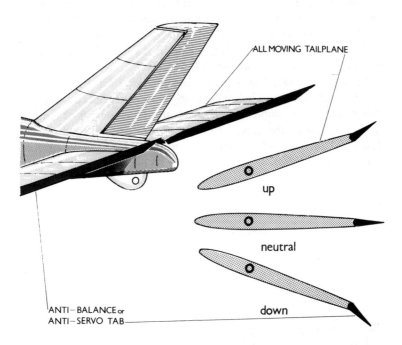

Fig. 89   All-moving Tailplane (known in the USA as a Stabilator). Note that the anti-balance tab is arranged to *add* control loads and prevent over-balance when the main surface is moved by the pilot.

linked to the control system that it moves in the *same* direction as the main control (the opposite direction to a balance tab). In this way it counters the adverse movement of the centre of pressure of the all-moving tailplane.

An advantage claimed for this type of control is that it allows more flexible loading of the cabin without incurring centre of gravity problems. However, some of the light aircraft with the longest cabins are, in fact, fitted with a fixed tailplane and separate elevators.

In the main, all-moving tailplanes do not handle as well as elevators.

Trim Controls

A pilot's attention during flight is divided among many things. There is the need to maintain a good lookout, steer an accurate heading, monitor the various flight and engine instruments and operate the radio.

If, because there was no load on the back seats, it was necessary to correct a nose-heavy situation by holding back the wheel/stick, yet another function would be imposed on the pilot and some of the other requirements would suffer. In any case, it is very tiring to hold on control pressure over a long period of time.

On larger aircraft trim controls are provided on the elevators, ailerons and rudder but in light training aeroplanes only an elevator trim is usual. The simplest form of elevator trim consists of tension springs linked to the elevator circuit. These are adjusted on the elevator trim control to remove out-of-balance loads on the wheel/stick (Fig. 90).

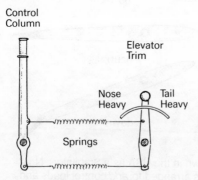

Fig. 90   Spring-loaded trim control, an old method which has been retained in several light aircraft of modern design.

Trim Tabs

The most common method of trim, one that may be used for all the primary controls in larger aircraft, is the **Trim Tab**, a small surface inset at the trailing edge of the elevators, ailerons or rudder. Each tab (when more than one control is fitted with adjustable trim) has its own trim control in the cabin.

Trim tabs move in the opposite direction to the required movement of the main control. For example, when there is a need to hold on prolonged UP elevator (i.e. wheel/stick back) the trim control in the cabin is moved back to lower the trim tab. This in turn applies a lifting force at the trailing edge of the elevators and removes the unwanted control force (Fig. 91).

When an all-moving tailplane is fitted it is the usual practice to incorporate trim capability within the anti-balance tab. In effect, adjustment of the trim control alters the position where the tailplane must be for the anti-balance tab to adopt a neutral setting relative to the trailing edge of the main control.

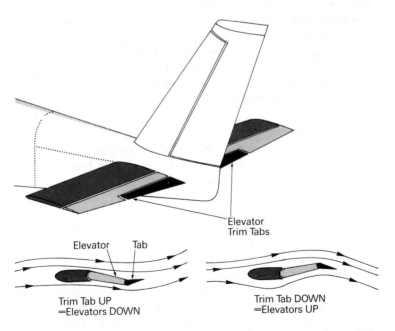

Fig. 91   Trim tab. The tab is adjusted by the pilot on a separate control.

Aileron Drag

Particularly at low airspeeds, sudden application of aileron will cause a yaw in the opposite direction to bank. This is more noticeable on light aircraft of early design and, while the characteristic is less conspicuous on modern types, it is important in so far as the ailerons tend to oppose turning unless they are of good design. In the case of a bank to the left the nose will yaw to the right because of **Aileron Drag**. The explanation is that, in order to bank, the aileron on the rising wing must be depressed into high-pressure air, at the same time increasing its angle of attack. Conversely the aileron on the down-going wing will be raised into lower pressure and its angle of attack will decrease. In other words the aileron on the up-going wing will have more drag than that on the lower wing so causing the aircraft to yaw momentarily towards the raised wing until overcome by the weathercock stability of the aircraft which results from 'further effects of aileron'.

Drag can be more evenly balanced between the two ailerons by arranging for the up-going one to move through a greater angle than the down-going surface. This is accomplished by simple geometric arrangement of the control linkage and aircraft fitted with the refinement are said to have **Differential Ailerons**.

An additional method is illustrated in Fig. 92 and it is often used in conjunction with differential ailerons. The hinge line is moved back behind the leading edge of the aileron and the hinge pins are carried on extensions from the wing so that the down-going aileron is in smooth continuance with the wings thus minimizing drag. Conversely, the up-going aileron protrudes its leading edge below the under-surface of the wing so causing additional drag on the inside of the turn. The leading edge of the aileron is often arranged to fit into a recess in the wing called a **Shroud** which further increases the efficiency of this type of aileron which is named **Frise** after its designer.

Frise ailerons have two further advantages:

1. The portion of the control surface forward of the hinge line may house the mass balancing weight (Fig. 93).
2. When unshrouded the area in front of the hinge line gives aerodynamic balance to assisting operation by the pilot.

Using both methods the results are very effective and the controls are then referred to as **Differential Frise Ailerons**. Unfortunately Frise ailerons are unsuitable for very high-speed aircraft because of the under-wing disturbance caused by the up-going surface.

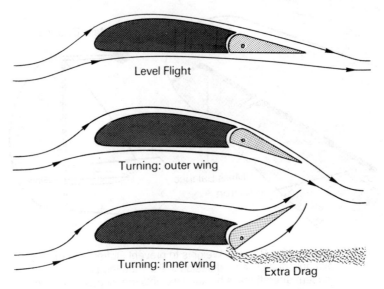

Fig. 92   Frise Ailerons. Centre and lower drawings show the position of the ailerons during a turn to the left.

Flutter (Fig. 93)

Modern materials have made possible the design and manufacture of extremely light airframe components of great strength. It is relatively simple to cater for the various stress requirements so that wings, fuselage, tailplane, etc., are strong enough to withstand loads in flight and landing but lightness means economy of material which is to the detriment of a further requirement – **Rigidity**.

Unless, for example, the wings possess rigidity, flexing may occur when the ailerons are operated, so twisting the wing and causing an adverse change in angle of attack on each wing when the function of the ailerons will be opposed or in extreme cases overcome. Alternatively the flexing of the wing may set up a violent oscillation or vibration between wing and aileron which if allowed to continue would within seconds result in structural failure. Such behaviour is not confined to the ailerons and **Flutter** may occur with the rudder or the elevators. Its cure may on occasions necessitate stiffening the affected structure, but the method in common use is called **Mass Balancing** when a weight is attached to the control surface, ahead of its hinge line, usually within the horn balance. The amount of weight

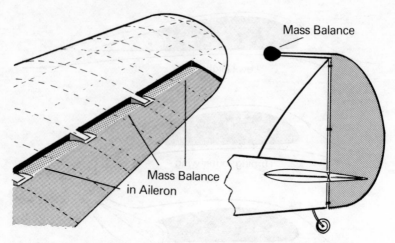

Fig. 93   Use of Mass Balance Weights to prevent Flutter. The rudder example depicts an old design to illustrate the principle. In modern aircraft mass balance weights are located in the horn area shown in Fig. 88.

is carefully calculated and high-speed aircraft are particularly sensitive to changes in mass balance to the extent that an extra coat of paint on the rudder may provoke flutter. It was for this reason that, many years ago, the red, white and blue stripes were transferred from rudder to fin on RAF aircraft as their performance improved.

Flaps, types and function

Figure 68 on page 141 illustrates four different airfoils and the point was made that, in general, thin wing sections are used for high speed while thicker airfoils are needed to generate enough lift when the aircraft is designed for relatively low speeds. It should be understood that no one airfoil section can be all things to all men; wings designed for high speed are not very good at flying slowly and *vice versa*.

Although much attention is devoted to reducing drag there is one phase of flight when a very efficient airframe with excellent gliding capabilities can be an embarrassment – the approach to land. The section headed Lift/Drag Ratio (p. 162) explained that the greater the amount of lift in relation to drag the better the glide performance in terms of distance covered per unit of height loss.

During the approach to land the pilot needs:

(*a*) As low a speed as possible consistent with safety;

(*b*) An approach path steep enough to provide good obstacle clearance; too flat a glidepath, the result of high L/D ratio, would not provide this. Also because of the almost level attitude of the aircraft forward visibility could, in some aircraft, be restricted.

Flaps achieve these aims by:

(*a*) Turning a relatively high speed airfoil into one capable of producing adequate lift at lower speeds. By altering the camber of a wing, flaps lower its stalling speed.

(*b*) As flaps are lowered they alter the L/D ratio, reducing the flaps-up figure by adding drag and so steepening the glidepath.

There are a number of different types of flap and these will shortly be described but, in general terms, maximum lift increase occurs during the first 15–25° of flap depression. This is accompanied by a smaller increase in drag. Further flap depression effects a small increase in lift and a larger increase in drag (Fig. 94).

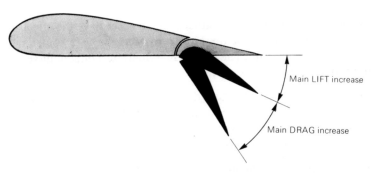

Fig. 94   With most flaps the main lift increase occurs during the first 15–20 of depression (according to design). Further depression of flap causes a modest lift increase along with a larger increase in drag.

Effect of Flap on Stalling Speed

The effectiveness or otherwise of flaps depends on their design but even the best of them can only reduce the stalling speed by a percentage of the flaps-up figure. Those fitted to light single-engine

171

trainers and tourers will reduce the stalling speed by 6 to 12 kt while the complex, high lift devices that are so essential to a large passenger jet can make a difference of 60 kt or more.

When flying at reduced speed it is the practice to lower part flap, usually 15–20°, for the purpose of reducing the stalling speed and lowering the nose from what could otherwise be a tail-down attitude at low speed, one that obscures the vision ahead. There is also the added bonus of having to use more power to balance the extra drag caused by the flaps. The increased slipstream that results from opening the throttle moves down the fuselage, over the tail surfaces and increases the effectiveness of the rudder and elevators.

In some, but by no means all, aircraft a part flap setting is recommended for take-off. In these cases flaps reduce the take-off roll by allowing lift-off at a lower speed. This technique, which is essential to high performance aircraft, has little effect on most light aircraft unless special short field take-off techniques are employed. These are described in Exercise 12, *Flight Briefing for Pilots, Vol. 1.*

Types of Flap

The following descriptions should be read in conjunction with Fig. 95.

Simple or Plain Flap. These rather crude devices have the effect of increasing the camber of the airfoil. In fact, they are sometimes known as **Camber Flaps.** They have a marginal effect on stalling speed and provide little increase in drag.

Split Flaps. Like the previously described flaps these are among the earliest designs that were introduced when the value of high-lift devices was recognized. They provide a slightly better lift increase but as drag producers split flaps are particularly effective.

Slotted Flaps. By arranging for the flaps to be hinged on short arms projecting below the wing surface (rather like a pair of inverted scissors with one blade tip fixed to the wing and the other attached to the flap leading edge) a gap or slot is made to open as the flaps are lowered.

The slot allows high pressure air from below the wing to spill through when it is induced to flow over the top of the flap, so preventing it from stalling when it it lowered to beyond the usual stalling angle.

Slotted flaps are very widely used. They provide a good lift increase but rather less drag than split flaps.

Fowler Flaps. These very efficient flaps lower the stalling speed by increasing the effective camber and adding to the wing area.

During the initial stages of lowering the flaps move back on tracks fixed to the wing structure, adding area and depressing slightly. Final movement is more or less confined to lowering of the flaps for the purpose of increasing drag. These very effective flaps are fitted to a number of iight aircraft and they form the basis of high lift devices used on large passenger jets.

The last two flaps in the illustration are developments of the basic Fowler flap design. They are not fitted to light aircraft but these complex devices may be seen on most of the passenger jets.

Flap operation and Limiting Speed

The mechanical construction of some of the flaps mentioned in the previous section is illustrated in Chapter 5, Airframes. Flaps may be operated via a simple mechanical linkage in which case the pilot will have a flap lever moving in a quadrant provided with indents at a number of settings. Apart from UP and DOWN the flaps may usually be set to a take-off position and one or two more intermediate positions before they are fully lowered.

Many light aircraft have electrically operated flaps and larger designs sometimes use hydraulic power. These will be described in Chapter 7, Aircraft Systems.

Whatever the type of flap, and however it is operated, there will be a **Maximum Flap Extension Speed**. This is listed in the aircraft manual and known in the 'V' code as V_{fe}. The airspeed indicator is marked with a white sector and the flaps may safely be lowered at any time when the needle of the ASI is within that speed range. If V_{fe} is exceeded structural damage could occur.

Slats, Slots and Spoilers

If an aircraft tends to drop a wing during the stall (i.e. a roll develops as the nose goes down) the down-going wing will be influenced by an airflow that meets it from below, thus effectively increasing the angle of attack and still further stalling that wing. Conversely, the up-going wing will meet its airflow from above, its angle of attack will be decreased and it will partly un-stall.

The situation is that one wing will be fully stalled while the other is flying. Remembering that drag increases after a wing has stalled the down-going, fully stalled wing will have more drag than the rising,

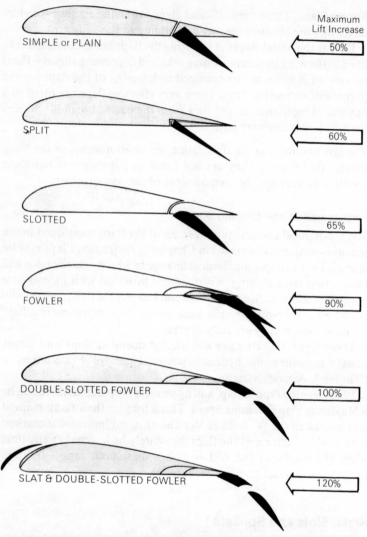

Fig. 95   Some of the more widely used types (see text). The two lowest examples are confined to larger aircraft.

partly un-stalled one. A yaw towards the dropped wing will be encouraged and all the ingredients then exist for a spin to develop.

To prevent the conditions just described some aircraft are fitted with **Auto Slats**, small airfoils which, in normal flight, fit closely around the wing leading edge. At high angles of attack the centre of pressure of the slat moves forward, pulling it away from the wing leading edge and opening up a slot through which high pressure air from below is directed to blow over the top surface, so maintaining what would otherwise be turbulent airflow leading to a stall. When the airspeed is increased again the slat's centre of pressure moves back and the auto slat closes with the wing (Fig. 96).

An alternative, non-mechanical method of preventing wing drop and delaying the stall is the use of fixed **Slots**. These are built into the wing surface opposite the ailerons and near the leading edge.

Well designed slats or slots can delay the usual 16° or so stalling angle and allow the wing to fly at 25° or more.

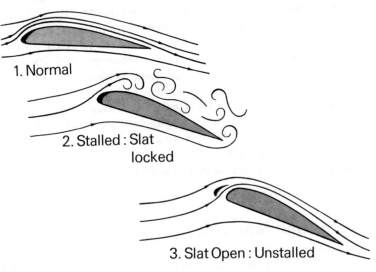

1. Normal

2. Stalled : Slat locked

3. Slat Open : Unstalled

Fig. 96 The Automatic Slat opens under the influence of the forward movement of its centre of pressure as the stalling angle is approached. Some auto slats may be locked closed for spinning and aerobatics.

Spoilers and Washout

A simple means of preventing wing drop at the stall is the **root spoiler**. These are sometimes known as **Buffet Strips** or **Buffet Inducers**.

The root spoiler takes the form of small strips, triangular in section and perhaps 12–18 inches long, which are affixed to the wing leading edges near where the wing joins the fuselage (Fig. 25, p. 116, Volume 1).

At high angles of attack the airflow breaks down around the strips, partly destroying the lift over the top surface of those areas of the wing that are adjacent to the fuselage. The effect of this is twofold:

1. The disturbed, turbulent air from the inner portions of the wing flows back over the tailplane where it impinges on the elevators to cause a warning buffet which can be felt through the elevators via the wheel/stick. It provides a good stall warning because the buffet occurs several degrees before stalling angle is reached.
2. Although the inner portions of the wing are stalled the outer areas remain unstalled and so remain flying, thus keeping the wings level.

The disadvantage of using root spoilers is that they cause the wing to stall at a higher speed than is really necessary. Also the **Pre-Stall Buffet** is only at its most effective when Root Spoilers are fitted to low-wing aircraft. In most high-wing designs the turbulent airflow created by the strips tends to pass above the tailplane and elevators.

A better method is the long established **Washout**. In most light aircraft the wing chord line makes an angle with the fuselage datum of 2–4°. This is known as the **Angle of Incidence**. A wing designed to have washout has a progressively reducing angle of incidence towards the wingtips so that when the inner areas of the wing have reached stalling angle the outer portions are unstalled and the tendency to drop a wing is minimized or even eliminated.

Forces and Moments

When wings, tail surfaces and the fuselage are put together and the controls have been installed the addition of an engine, propeller and instruments completes the aeroplane. While the private pilot is not expected to be an aerodynamicist it is important that there should be an elementary knowledge of the forces being imposed in flight. Such knowledge can prevent accidents that may result from attempting to fly the aircraft outside its capabilities.

Balanced flight – straight and level

Taking straight and level flight as the starting point two pairs of forces should be considered.

Assuming the aircraft weighs 1500 lb the same amount of lift will be required if it is to maintain height. The term centre of pressure has already figured in this chapter. To understand it more fully imagine a model aeroplane hanging from the ceiling by a thread attached to each wing. Point of attachment of these threads may be regarded as the centre of pressure.

The aircraft, complete with fuel, occupants and baggage, has a centre of gravity. A degree of latitude is allowed in determining this because there is a **C of G Range** which may extend for 4–6 in. in a light trainer and 12 in. in the case of a light twin-engine design. The subject is dealt with more fully in Chapter 8.

In the interests of longitudinal stability the two forces, lift pulling up and weight pulling down are arranged so that, instead of being in line (i.e. centre of pressure directly above the centre of gravity) weight is slightly ahead of lift. Study Fig. 97 and regard the two arrows, LIFT and WEIGHT as lengths of string pulling in the direction shown. Because the two forces are not in line there will be a nose-down **Couple**, a tendency for the aircraft to be nose heavy which will cause our model aircraft to hang from the ceiling in a slightly nose-down attitude. In flight the nose-down couple is partly balanced by a down load on the tailplane/elevators and such an arrangement will provide longitudinal stability.

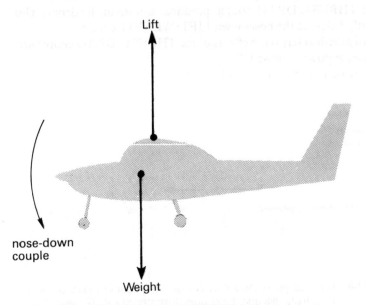

Fig. 97   Nose-down pitch caused by Lift/Weight couple.

For the aircraft to cruise at a steady speed thrust from the propeller must equal drag. Some student pilots find this difficult to understand because, on the face of it, when a force pulling forward is balanced by an equal force pulling back the vehicle (car, ship or aircraft) should stand still, rather like two perfectly matched tug-of-war teams. Look at it this way.

Aircraft X is about to take off. The pilot opens the throttle and immediately the propeller generates 200 lb of thrust. The brakes are on, the aircraft is stationary so there is no drag. Acceleration starts as soon as the brakes are released and as the speed increases so does the drag – 50 lb, 120 lb, 160 lb until at 200 lb drag balances thrust and there is no further gain in speed which, by then may be over 100 kt (the aircraft will, of course have taken off and been placed in the straight and level attitude).

If the pilot reduces power so that thrust becomes 150 lb, drag, being in excess of thrust, will slow the aircraft. As it reduces speed so drag decreases until it becomes 150 lb when the aircraft will settle at the new speed with thrust and drag in balance.

The aircraft designer aims to arrange thrust so that its line of force is below drag. In Fig. 98 it will be seen that the THRUST arrow is below that for total DRAG. If the two arrows are again regarded as lengths of string pulling in the directions indicated it will be clear that the THRUST/DRAG couple produces a nose-up tendency. This partly balances the nose-down LIFT/WEIGHT couple.

In practical terms the effects of the THRUST/DRAG couple are:
Increase power = nose UP
Decrease power = nose DOWN

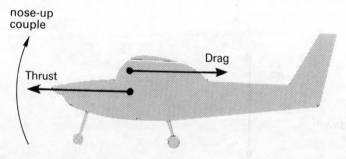

Fig. 98   Nose-up pitch caused by Thrust/Drag couple. In straight and level flight this and the couple illustrated in Fig. 97 are in near-balance.

One of the advantages of this arrangement of forces is that if there is an engine failure the aircraft will automatically lower its nose and move towards the gliding attitude.

Manoeuvres

Through a combination of control inputs via the wheel/stick, rudder pedals and the throttle, an aircraft can be made to perform a number of manoeuvres, the more complex being aerobatics (loops, rolls, etc.) but these are outside the scope of this book and not included in the PPL syllabus.

Manoeuvres of vital, every-day importance to the pilot are:

Straight and level flight
Climbing
Descending
Turning
Stalling
Spinning

The aerodynamics and forces involved in these manoeuvres are explained in the handling manual, *Flight Briefing for Pilots Vol. 1*, Exercises 6, 7, 8, 9, 10A, 10B and 11. To avoid duplication of text this chapter on principles of flight ends here.

Questions

1 **When air is induced to flow over an airfoil section:**
   (a) Pressure is reduced over the top surface,
   (b) Pressure is increased over the top surface,
   (c) Pressure is reduced over the top surface and increased below the lower surface.

2 **The centre of pressure is:**
   (a) The point through which the total effect of lift may be said to act,
   (b) The point at which maximum drag occurs,
   (c) The force oposing the centre of gravity.

3 **The angle of attack is:**
   (a) The angle between the airfoil chord line and the relative airflow,
   (b) The angle between the relative airflow and the angle of incidence of the wing,
   (c) A constant determined by the manufacturers.

4   Principles of Flight

4   **For a given airspeed lift increases:**
   (a) As the angle of attack is increased,
   (b) As the centre of pressure moves forward,
   (c) As the angle of attack increases and the centre of pressure remains constant.

5   **As the angle of attack is increased:**
   (a) The drag is reduced,
   (b) The drag remains the same if the speed is unchanged,
   (c) The drag is increased.

6   **While performing its stabilizing function the tailplane:**
   (a) Produces no lift,
   (b) Produces lift,
   (c) Produces a correcting force, either up or down.

7   **Directional stability achieved by the fin is also influenced by:**
   (a) The keel surface or area behind the centre of gravity,
   (b) Movement of the centre of pressure,
   (c) Using the 'high wing' design.

8   **If the profile drag at 100 kt were found to be 200 lb, the drag at 200 kt would be:**
   (a)   400 lb,
   (b)   800 lb,
   (c) 1200 lb.

9   **The primary effect of aileron is to cause movement in the rolling plane. The further or secondary effect is:**
   (a) Yaw followed by a spiral dive,
   (b) A skid outwards,
   (c) A nose-up attitude as a result of increased lift on the up-going wing.

10   **The further effect of rudder causes:**
   (a) The aircraft to slip,
   (b) A movement in the rolling plane,
   (c) A movement in the rolling plane followed by a spiral dive.

180

11 When the elevator trim tab is set in the up position this will:
    (a) Assist the pilot to maintain a nose-up attitude
    (b) Will slow down the airflow over the control surface and make it more effective,
    (c) Assist the pilot to maintain a nose-down attitude.

12 When an aircraft is flying straight and level at a constant IAS the forces are as follows:
    (a) Lift and weight are equal and drag is proportional to the relative airflow,
    (b) Lift and weight are equal and thrust is proportional to the relative airflow,
    (c) Lift and weight components are equal and thrust is equal to drag.

13 When power is adjusted in level flight the slipstream will tend to affect:
    (a) The longitudinal stability,
    (b) Directional stability,
    (c) Lateral stability.

14 The maximum rate of climb is achieved at:
    (a) A low airspeed and a high power setting,
    (b) A high airspeed and a high power setting,
    (c) A compromise between speed and power setting.

15 Flaps are fitted to an aircraft for the purpose of:
    (a) Increasing lift and drag while decreasing the stalling speed,
    (b) Increasing drag, increasing payload and increasing the stalling angle,
    (c) Increasing lift, drag and stalling speed.

16 In adddition to the usual benefits, a Fowler flap:
    (a) Provides a pre-stall buffet,
    (b) Increases the wing area,
    (c) May be used to improve the take-off.

17 In a correctly executed rate 1 turn an aircraft will change direction at:
    (a) 360° per min,
    (b) 2° per sec,
    (c) 3° per sec.

4 Principles of Flight

18 During a turn total lift balances weight and:
  (a) Accelerates the aircraft towards the centre of the turn,
  (b) Balances weight and does not affect the turn,
  (c) Has no turning force, the increased lift resulting from the outer wing travelling faster than the inner wing during the turn.

19 The rate of turn is dependent on:
  (a) The airspeed and the angle of bank,
  (b) The airspeed and the angle of attack,
  (c) The angle of bank and the power available.

20 Which of these statements is correct?
  (a) Stalling can only occur in certain attitudes,
  (b) Stalling can occur in almost any attitude,
  (c) Stalling can only occur when the airspeed is low.

21 The anti-servo tab on an all-flying tailplane:
  (a) Moves in the same direction as the main surface to assist the pilot by removing control loads,
  (b) Moves in the opposite direction to the main surface to assist the pilot by removing control loads,
  (c) Moves in the same direction as the main surface to assist the pilot by adding control loads.

22 During a spin an aircraft is simultaneously:
  (a) Pitching up, yawing and rolling,
  (b) Pitching down, yawing and turning,
  (c) Pitching up and down, yawing to the accompaniment of severe slip towards the spin axis.

23 To recover from a spin the aircraft must be made to:
  (a) Stop rolling with aileron and attain level flight with power and elevator.
  (b) Stop the roll with opposite rudder and attain level flight with power and elevator,
  (c) Close the throttle, ailerons neutral, stop yawing with opposite rudder, decrease its angle of attack with forward wheel/stick, centralize the rudder after spinning stops and then resume level flight with correct attitude and power.

24 There are four possible causes of swing during a take-off. Those affecting a nosewheel aircraft are:
  (a) Torque effect and slipstream effect,
  (b) Slipstream effect, gyroscopic effect and torque effect,
  (c) Asymmetric blade effect, and torque effect.

25 During a short take-off, use of the recommended flap setting will:
  (a) Decrease the take-off run and increase the rate of climb,
  (b) Decrease the take-off run, increase the climb angle and probably reduce the rate of climb,
  (c) Decrease the take-off run without affecting the rate of climb.

26 In a turn at a steep angle of bank:
  (a) The stalling speed is increased because the angle of attack must be increased,
  (b) The stalling speed is increased because the wing loading is increased,
  (c) The stalling speed is increased because of the inclined lift.

27 Mass balance is fitted to some controls for the purpose of:
  (a) Assisting the pilot to move a heavy control surface,
  (b) Preventing flutter,
  (c) Opposing aerodynamic loads during high 'g' manoeuvres.

28 Lateral stability is built into an aircraft by incorporating some of the following features:
  (a) High wing, dihedral angle, sweepback, high keel surface,
  (b) Dihedral angle, washout, Frise ailerons,
  (c) Dihedral angle, washout and slats.

29 Some control surfaces are fitted with horn balance for the purpose of:
  (a) Preventing flutter,
  (b) Preventing aileron drag,
  (c) Relieving the pilot of otherwise heavy control loads.

30 On some aircraft small strips are fixed to the leading edges of the wings, close to the fuselage. Their purpose is to:
  (a) Make the aircraft stall more cleanly,
  (b) Prevent a wing dropping at the stall,
  (c) Prevent a wing dropping at the stall and provide a pre-stall warning in the form of a buffet.

31 **When a wing drops during stalling practice the ailerons must not be used to regain lateral level because:**
   (a) The down-going aileron on the wing to be raised would be more fully stalled causing an increase in drag and the risk of a spin,
   (b) At low speeds near the stall the ailerons are not very effective,
   (c) The up-going aileron on the wing to be raised would aggravate the situation.

32 **It is potentially dangerous to commence a gliding turn at a low airspeed because:**
   (a) The risk of stalling is greater in a turn,
   (b) At low speeds there is a risk of aileron reversal,
   (c) There is a risk of an incipient spin developing.

33 **A wing is most efficient when it is flown at an angle of attack of $3\frac{1}{2}°$–$4°$. This is called:**
   (a) The riggers angle of incidence,
   (b) Minimum drag angle,
   (c) Best lift/drag angle.

34 **Most aircraft have a tailplane. What is its purpose?**
   (a) To provide longitudinal stability,
   (b) To carry the elevators,
   (c) To cater for changes in weight and balance.

35 **What is wheelbarrowing?**
   (a) The tendency to pitch forward during landing due to harsh application of the brakes,
   (b) The tendency during take-off or landing for the aircraft to run along on the nosewheel with the main undercarriage off the ground,
   (c) Instability on the ground due to a faulty nosewheel steering damper.

Chapter 5
Airframes

Recent experience has emphasized that some accidents could have been avoided had the pilot understood a little more about the aeroplane he was flying. It seems strange that so many pilots (some of them professional) should have little knowledge of airframes and this chapter is intended to deal with the subject in general terms.

Structures

Given the right materials most constructional problems can be solved by good engineering principles. The Romans, limited in the main to the use of stone, contrived fine buildings and aqueducts that, to this day, remain a monument to their skill and imagination. Jumping the centuries, iron and later steel made possible great bridges and such eccentricities as the Eiffel Tower. In all of these examples the problem of strength was dealt with simply by using plenty of stone, iron or steel. Weight was of little consequence. The aeroplane presented a completely new set of problems to its constructors, many of whom had no formal or even *ad hoc* engineering training and aerodynamics apart, the most difficult of these problems was the need to provide strength and lightness. Much ingenuity was and still is required to cater for these conflicting requirements. Since many aircraft of old design continue to fly it is appropriate to outline some of the methods used to achieve strength with minimum weight.

Fuselage construction

The purpose of the fuselage is to house the payload and provide a lever or moment arm through which the tail surfaces can act (see pages 143–5, Chapter 4).

The early aircraft almost invariably had an open fuselage comprised of three of four wooden **Longerons** which ran the length of the aircraft. They were held apart by a number of struts to form a long box, the entire structure being braced with piano wire under tension which gave it rigidity (Fig. 99). Later, wood began to be replaced by steel tube, typical examples of this type of construction being the Bristol Bulldog and the Gloster Gauntlet. A further development dispensed with the bracing wires and embodied diagonal bracing members, the entire structure being welded. To this primary frame was added a light secondary structure of **Stringers** for the purpose of supporting the doped fabric covering and providing a better aerodynamic shape. The welded steel tube fuselage was embodied in the Auster and early Piper series of light aircraft, the Tiger Moth and indeed it is used in a number of modern designs of simple concept. This form of construction has the advantages of great strength, relative cheapness and ease of repair, the disadvantage being the need for a secondary structure of stringers or removable panels to provide good aerodynamic shape (Fig. 100).

In 1912 the French designer Ruchonnet dispensed with the then universal practice of using longerons, struts and wire bracing. His Deperdussin monoplane (for several years holder of the world's air speed record) featured a fuselage of **Monocoque** construction, a system based upon the eggshell concept whereby an unbroken box, for preference, with curved sides, is possessed of great strength, rigidity and lightness. A true monocoque structure is without frames or longerons, all constructional material being confined to the walls or skin. Unfortunately a structure of this kind is weakened by the introduction of openings such as those needed for entrance to the cabin, cabin windows, attachment of the wings and tail surfaces etc., therefore a modified technique has been developed which offers most of the advantages of monocoque construction while allowing the designer more freedom in providing for essential openings. A light structure of longerons and frames is covered with a skin of sufficient strength to cater for bending loads resulting from the tail surfaces, twisting loads and the various stresses that occur while manoeuvring on the ground or encountering turbulence in the air. Since the only purpose of the inner structure is to provide shape to the outer skin and perhaps local strengthening for the attachment of the tailplane, wings, etc., sometimes the longerons are replaced by a number of light stringers. Such a structure is known as **Stressed Skin**. It may be of metal or wood and it is the most common method of construction currently used for large and small aircraft alike. Metal structures are usually of riveted aluminium alloy while wooden airframes are most commonly of spruce framing to which is glued an outer skin of high-quality beech ply, the entire surface being covered by a fine

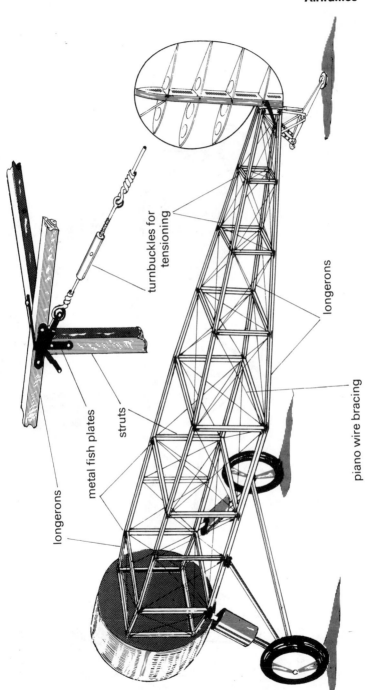

turnbuckles for
tensioning

longerons

piano wire bracing

struts

metal fish plates

longerons

Fig. 99   Typical wire-braced wooden fuselage of the 1912–1919 period. Later wire-braced aircraft were constructed of metal tubes.

187

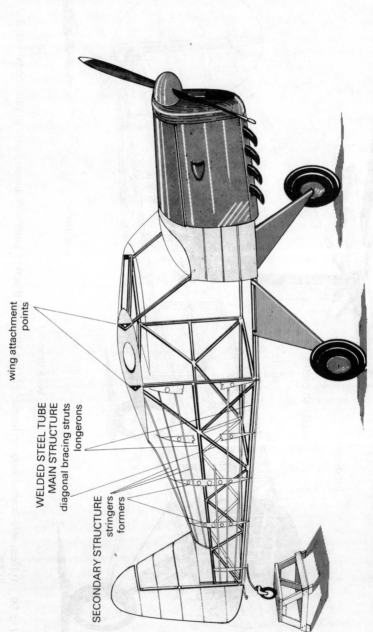

wing attachment
points

WELDED STEEL TUBE
MAIN STRUCTURE
diagonal bracing struts
longerons

SECONDARY STRUCTURE
stringers
formers

Fig 100. Welded steel tube fuselage of a type that is still employed in some modern light aircraft. Note the absence of bracing wires which are replaced by diagonal tubes. This type of fuselage is known as 'Warren Girder' construction.

fabric which protects the outer surface of the ply and provides a good foundation for the paint finish. Modern synthetic glues are very durable, being unaffected by age or damp. A typical metal stressed-skin fuselage is shown in Fig. 101.

In more recent times there has been another development embodying two skins of glass-fibre separated by a honeycomb layer called Nomex which is some ¼ in. thick. The glass-fibre/Nomex sandwich is wrapped around a full size plaster mould of the fuselage, each layer having been pre-impregnated with resin adhesive, and the assembly is then placed in an oven. Vacuum is introduced to ensure proper flow of the resin and good contact between layers, then the sandwich is baked at 100°C.

What emerges is a fuselage of great rigidity, light weight and near perfect surface.

Wing construction

The purpose of the wing is to generate lift and to support the weight of the aircraft. The first function is achieved by providing the wing with **Ribs** of airfoil shape (see page 134, Chapter 4). In aircraft of early design the weight of the aircraft was supported on one or more **Spars** over which were positioned a number of ribs of airfoil section, the entire structure being fabric covered (Fig. 102). Most early aircraft were biplanes with relatively thin wings prevented from folding under the weight of the aircraft by a system of **Bracing Wires** stretched diagonally between the upper and lower wings. While this arrangement provided a light structure of great strength the braced biplane suffered from high drag so that many designers turned towards the monoplane, at first with thin **Strut-Braced** wings (Puss Moth, Piper Cub and the Auster series). Later designs dispensed with all external bracing and so reduced drag to a minimum. Stiffness sufficient to carry the weight of the aircraft is catered for by thickening the wing, so allowing for spars of greater depth. This is called a **Cantilever** wing. The various methods of achieving wing rigidity are illustrated in Fig. 103. Cantilever wings may be constructed of wood or metal. They may feature one or more spars of substantial proportions and a series of ribs, the entire structure being fabric covered. More usually the wing will be of the stressed-skin type with lighter spars, fewer ribs and a number of stringers to support the skin which itself contributes to the strength of the wings and resists twisting. Such a wing is illustrated in Fig. 104. Although this is the most common type of wing in current use a combination of methods is sometimes incorporated

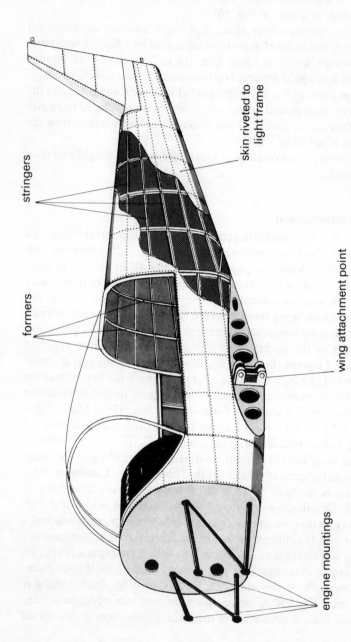

stringers

formers

skin riveted to light frame

wing attachment point

engine mountings

Fig. 101 Modern stressed skin fuselage. Note the absence of heavy structural members, many of the loads being absorbed by the outer skin which is supported on light frames and stringers.

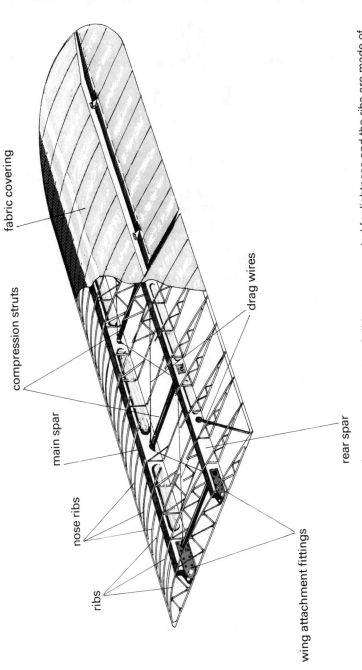

fabric covering

compression struts

main spar

nose ribs

ribs

drag wires

rear spar

wing attachment fittings

Fig. 102 Wire-braced biplane wing of wooden construction. Solid spars are routed for lightness and the ribs are made of wood strips.

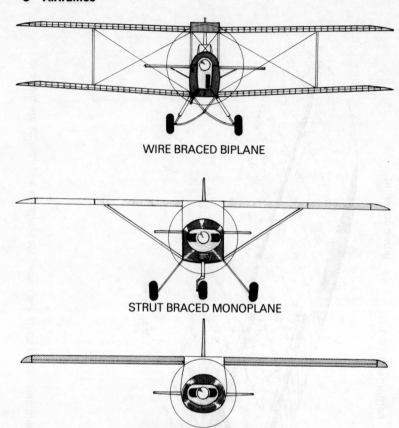

WIRE BRACED BIPLANE

STRUT BRACED MONOPLANE

CANTILEVER MONOPLANE
(high wing)

CANTILEVER MONOPLANE
(low wing)

Fig. 103 Four methods of attaining wing rigidity. In the two lower
examples external struts and bracing wires are replaced by
stronger and deeper wing spars. These are known as
Cantilever Wings.

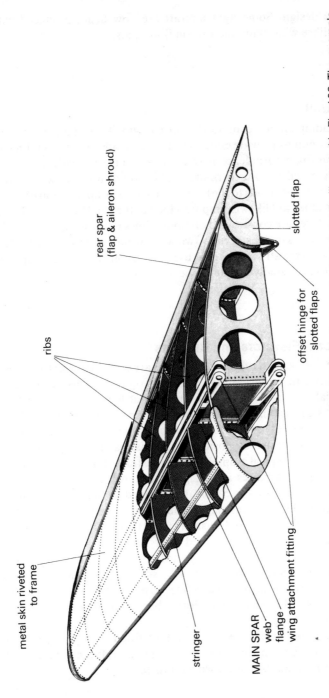

ribs

rear spar
(flap & aileron shroud)

slotted flap

offset hinge for
slotted flaps

metal skin riveted
to frame

stringer

MAIN SPAR
web
flange
wing attachment fitting

Fig. 104 Modern stressed skin wing. Compare its simplicity with the very detailed structure illustrated in Fig. 102. The example shown here is of light alloy but, like the stressed skin fuselage shown in Fig. 101 wood or composites may be used (e.g. glass fibre).

193

in the design. Some light aircraft are now being produced with glass-fibre wing skins and carbon fibre spars.

Tail Unit

Depending upon the aircraft's centre of gravity the tailplane will be called upon to provide either up or down loads, while from time to time in-flight turbulence may cause a change in loading on both tailplane and fin. Furthermore application of up or down elevator will add further loads to the fixed tailplane while application of left or right rudder will likewise impose a load to the fixed fin. Generally the tail surfaces are constructed on similar lines to the wings and may be of stressed skin construction (wood or metal) or internally braced, fabric covered structures. Ribs used in the construction of tailplanes and fins are usually of symmetrical section.

Ailerons, Elevators and Rudders

Since the elevators, rudder and ailerons are thin, tapering structures it is difficult to achieve a high degree of rigidity unless heavy gauge materials are used. There is a tendency at high airspeeds for these surfaces to flex and then set up a vibration cycle known as flutter which can rapidly cause disintegration of the wings, fin or tailplane. Flutter is prevented by positioning a mass balance weight ahead of the hinge line, typical examples being illustrated on page 170, Chapter 4. All control surfaces are designed to move through a particular number of degrees and **Stops** are provided to limit the amount of movement. The surfaces may be connected to the pilot's controls via a system of **Push-Pull Rods** (usually tubes) running in roller guides or high-tensile steel cables guided over pulleys. Cable tension and control alignment may be adjusted through suitably positioned **Turnbuckles**. Some aircraft make use of both systems but the importance of ensuring full and free movement of all controls before take-off is self evident. A typical cable-operated control circuit is illustrated in Fig. 105.

Larger and faster types of aircraft are fitted with aerodynamic or power-assisted controls. In so far as light aircraft are concerned the student need not be concerned with these.

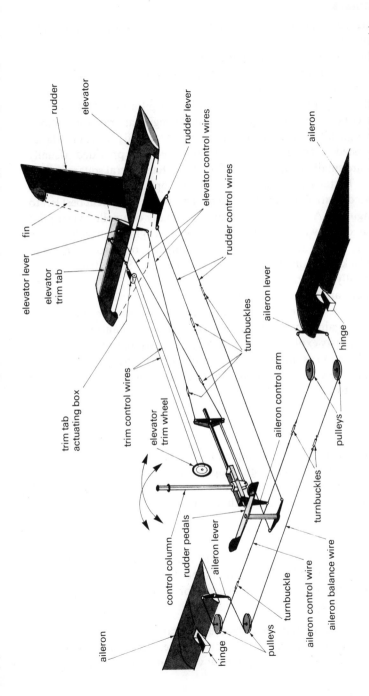

Fig. 105 Cable-operated control system showing location of levers, pulleys and turnbuckles used to attain the correct cable tension. In some aircraft pairs of cables are replaced by a single push-pull tube running in bearings.

Trimmers

In the broadest terms trimmers may be considered under two headings:

1. Those adjustable by the pilot in flight.
2. Those adjustable to a fixed setting by a mechanic while the aircraft is on the ground.

While larger types of aircraft usually have pilot-operated trimmers on all controls the majority of light single-engine aircraft are confined to an adjustable elevator trim only and aileron and rudder imbalance is adjusted on the ground by using the fixed tabs provided. Usually these are small metal surfaces of no more than several square inches' area. The tabs are bent slightly in the opposite direction to the required control surface movement. This is an adjustment best left to an engineer. Adjustable trim tabs are usually controlled via a pilot operated trim wheel or handle connected to a lightweight cable which in turn moves the trim tab, either directly or through a small screwjack positioned in the adjacent structure. Some trimmers are electrically operated.

When an all-moving tailplane is fitted the risk of control overbalance due to unstable movement of its centre of pressure is guarded against by the provision of an anti-servo or anti-balance tab. As previously explained in Chapter 4, this is arranged to move in the *same* direction as the tailplane so adding load to the control. The tab may also be adjusted relative to the tailplane by the pilot's elevator trim control.

Flaps

Flaps vary in type from simple hinged panels which are little more than drag producers to highly sophisticated double or treble slotted Fowler devices as seen on large jet aircraft (Fig. 95). Usually a flap follows the constructional method of the wing. It may be attached by:

1. A simple hinge in the case of simple or plain flaps and split flaps (Fig. 106).
2. An offset hinge that provides a small gap in the down position when the flap is of the slotted type (Fig. 107).
3. A system of runners and wheels for area extending flaps of the Fowler type (Fig. 108).

Actuation of the flaps may be mechanical via a lever and suitable linkage, provision being made for intermediate settings. However, electrically-operated flaps are now very common and these may in

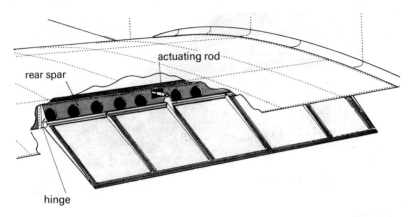

Fig. 106 Split flap.

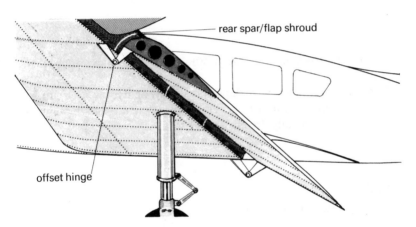

Fig. 107 Slotted flap showing off-set hinges that provide a small gap
between wing and flap as it is lowered.

most cases be set to any angle of depression within the range of flap
movement. Alternatively some aircraft have a three-position switch
providing the following flap settings:

1. UP
2. Optimum lift position
3. Full flap.

Probably the most convenient type of flap control is the **Follow-up**

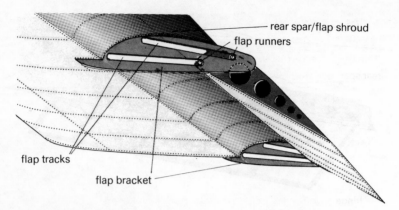

rear spar/flap shroud
flap runners
flap tracks
flap bracket

Fig. 108 Fowler flap (near section has been removed for clarity). Note
the tracking arrangement which enables the flap to increase
wing area by moving back during the early stages of lowering.

system; the flaps move to the position set on the flap switch, then
stop.

Certain aircraft types have flaps operated by an hydraulic circuit. In
these cases pressure may be generated by an engine-driven pump or
one that is electrically driven.

Slats

There are two main types of slat:

1. Fixed
2. Automatic

The fixed slat may be a small metal plate curved to follow the
contours of the wing leading edge. It is held in position by brackets so
that a slot is formed between the slat and the wing. Alternatively a
series of slots may be built into the leading edge of the wing, usually
ahead of the ailerons so that lateral control is maintained at high
angles of attack.

Automatic slats are mounted on a linkage that allows them to lie
flush with the wing leading edge when flying at cruising speed angles
of attack. Their aerodynamics and operation are described on page
173, Chapter 4.

Undercarriage

The obvious task of the undercarriage is to allow the aircraft to manoeuvre on the ground while taxying, taking-off and landing but it is the last named that makes the biggest demands on the undercarriage, particularly on training aircraft operating off rough grass surfaces. Clearly some form of springing must be incorporated and the most common methods used are:

1. Telescopic undercarriage legs which use coil springs, rubber in compression or compressed air as a spring medium. To minimize recoil after, say, a heavy landing some kind of damping is provided. In the case of rubber in compression or springs this often takes the form of a friction device arranged to expand between the sliding tubes as the leg extends after being compressed. Undercarriage struts using compressed air as a springing medium incorporate an oil damping system. Such undercarriages are known as **Oleo Struts**.

2. Leaf-spring undercarriage suspension where the wheels are carried on simple flexible legs which extend from the fuselage. They may be of spring steel or glass fibre.

Most undercarriages are now of the nose-wheel type and particular care must be exercised during the take-off and landing to ensure that all avoidable load is removed by applying back pressure to the elevator control in the manner prescribed in the handling chapters dealing with these manoeuvres in *Flight Briefing for Pilots, Vol. 1*.

Nosewheel Steering

While it is common practice to link the nosewheel to the rudder pedals, either directly or through springs, and so provide directional control while on the ground, some aircraft have free castoring nosewheels. Since the pilot has no direct control over a castoring nosewheel, directional control during taxying is dependent upon the use of differential brake and slipstream, while at higher speeds, e.g. during the take-off and initial landing run, the aerodynamic effect of the rudder will control direction.

The nosewheel may, at certain speeds tend to **Shimmy** (vibrate from side-to-side). The vibration can be of sufficient intensity to be unpleasant. To overcome this an **Anti-shimmy Damper** is usually fitted to the nosewheel strut. It takes the form of a small piston moving in an oil-damped cylinder.

When a tailwheel undercarriage is fitted (and some modern light aircraft do have tailwheels) the same arrangements apply as those described for nosewheels.

Whatever the type of undercarriage when it is fixed (i.e. non-retracting) considerable reductions in drag will result from careful design of wheel spats and leg fairings. Some manufacturers include a warning in the flight manual to the effect that removal of the wheel spats will markedly affect both rate of climb and speed. A typical streamlined undercarriage leg and wheel is illustrated in Fig. 109.

Brake Systems

Early aircraft relied entirely upon a tailskid to provide braking effect. While this somewhat crude device was fairly satisfactory on grass surfaces, in conditions of even light wind attempts to control, say, a Tiger Moth on a hard surface could be almost impossible, particularly when a crosswind exists.

Early brakes were of the expanding shoe type still used in many light family cars. These were cable operated via a hand lever or heel pedals. In the former case provision was usually made for differential braking via a linkage to the rudder pedals thus providing more braking effect on the inside of the turn when this was required.

The shortcoming of drum brakes, whether fitted to a car or aircraft is **Fade** — a progressive loss of braking efficiency that is proportional to the amount of brake being applied. During prolonged braking the shoes and drums become heated causing a loss of friction so that braking effort declines often at the very time when it is most needed. To overcome the problem **Disc** brakes were developed for aircraft and later introduced to higher performance cars. By confining the friction area to one or more pairs of **Pads** the disc (attached to the wheel) is exposed to cooling air so ensuring lower operating temperatures and greatly reduced fading.

In light aircraft the brakes are usually operated by a simple self-contained hydraulic circuit similar to that illustrated in Fig. 110. The pads, one of them fixed, the other moved by a single, hydraulic piston, must be replaced when wear reduces them to the minimum permitted thickness.

In most aircraft provision is made to lock on the brakes for parking. It should be understood, however, that unlike a car handbrake, an aircraft handbrake simply locks on the brakes. There is no separate handbrake circuit.

Before flight a pilot should check for obvious signs of leaking hydraulic fluid.

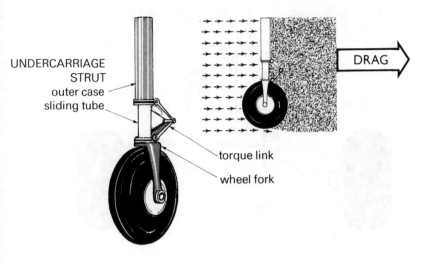

UNDERCARRIAGE
STRUT
outer case
sliding tube

DRAG

torque link

wheel fork

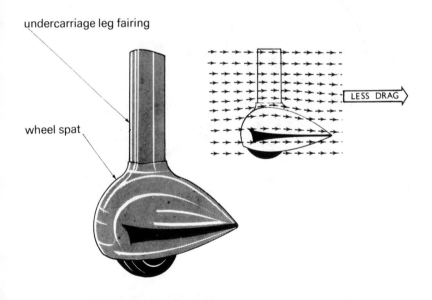

undercarriage leg fairing

wheel spat

LESS DRAG

Fig. 109 Reducing drag by streamlining. For light aircraft of moderate
power and performance this method is more suitable than the
weight, cost and complexity of a retractable undercarriage.

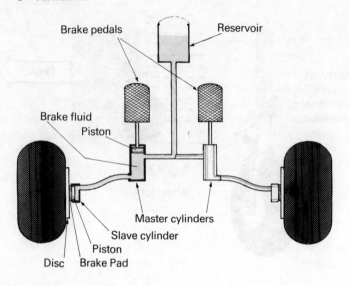

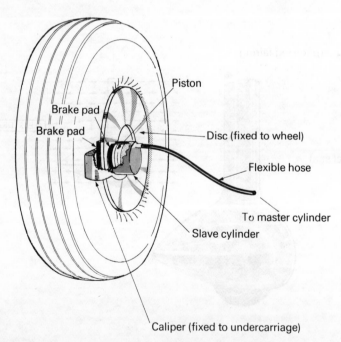

Fig. 110 Simplified illustration showing a hydraulic disc brake installation as fitted to most light aircraft. Lower drawing shows the two brake pads, one fixed behind the disc and the other moved under pressure by the slave cylinder/piston.

Seating

Unlike most vehicles the aeroplane is capable of generating 'g' which, in effect, increases the weight of both aircraft and occupants according to the loading imposed. Without resorting to aerobatics it is easily possible to subject an aircraft to considerable loading. In a 60° bank turn weight is doubled, while in a correctly executed 70° bank steep turn a 160 lb pilot would exert a force of 480 lb in his seat. Since an aircraft in the utility category is stressed for up to +4.4 g with a 50 per cent safety factor this means that the seat in a non-aerobatic category light aircraft must be able to support over six times the weight of the largest pilot likely to fly the machine.

Seats may be adjustable for rudder pedal reach by sliding back and forth on rails, the usual car-type locking catch being positioned under the front edge of the seat. There may in addition be a height adjustment. Some seats are fixed, height being catered for by adding cushions while the rudder pedals are themselves adjusted for reach.

Unless the aircraft is intended to be flown while wearing parachutes there will be no provision for these since the seat-type parachute must fit into a bucket seat intended for the purpose. The design of all seating, domestic, automobile or aircraft can often leave much to be desired in so far as correct support for the human frame is often overlooked. It is particularly important that pilots, experienced or under training, should ensure that the seats and/or pedals are correctly adjusted to allow full and free movement of all flying and engine controls.

Safety Harness

Safety straps are of many designs, the four- or five-strap type retained in a central quick release box being most satisfactory. Some aircraft are fitted with a lap strap to which is attached a single diagonal strap that passes over one shoulder. Sometimes this is of the inertia-reel type allowing freedom of movement until the body is jerked forward as would be the case in an accident. Generally diagonal straps are not well positioned to give comfortable protection. Furthermore there is a tendency with some designs for the metal fastening to become detached from the lap strap during flight. Pilots are therefore advised to exercise caution while wearing this type of harness.

Types and purposes of Control Locks

On a windy day the ailerons, elevators and, on aircraft without nose-wheel steering, the rudder may be blown against the control stops with sufficient violence to cause unnecessary strain and even damage. In the case of tailwheel aircraft it is not unknown for the tail to lift off the ground in gusty conditions. When there is no provision for locking the controls it is good practice to secure the control wheel or stick with the safety harness but most aircraft are now provided with a control lock or locks which may take the form of:

(a) External clamps that secure the ailerons to the wings, the elevators to the tailplane and the rudder to the fin.

(b) A pin which is inserted through the control wheel shaft so preventing movement of the ailerons and elevators. The rudder is usually restrained from moving because the rudder pedals are linked to the nosewheel.

(c) Other mechanical locks within the cabin.

The consequences of taking-off with the controls locked are obvious. Indeed there have in the past been fatal accidents of this kind. With a view to preventing such accidents external clamps usually have a red pennant attached, while locking pins often incorporate a red warning plate or flag which covers the ignition and master switch. Other types of mechanical lock are sometimes arranged to limit throttle movement so long as they are in position.

It is essential to remove all external clamps or interior control locks before commencing the pre-flight external check, a precaution that is backed up after entering the aircraft by a 'full and free movement' control check.

Questions

1 **Stressed skin aircraft are constructed as follows:**
   (a) Metal sheets riveted to steel tubes,
   (b) Outer sheets attached to a light structure of longerons and frames. The airframe may be of wood or metal,
   (c) Pressed aluminium alloy with no supporting frame.

2 **In a modern aircraft fitted with a cantilever wing weight is supported by:**
   (a) One or more spars of considerable depth,
   (b) A series of ribs,
   (c) Internal wire bracing.

3 **Flutter is caused by:**
(a) Turbulent airflow over the wings,
(b) Turbulent airflow over the wings and the tail surfaces,
(c) Vibration due to flexing of the ailerons, rudder and/or elevators.

4 **Some control surfaces carry a small metal tab on their trailing edge for the purpose of:**
(a) Providing trim adjustment by bending to the correct position,
(b) Indicating the manufacturer's part and release numbers,
(c) Discharging static electricity and improving radio reception.

5 **Flaps which not only depress but also increase the wing area by extending backwards on tracking are called:**
(a) Slotted flaps,
(b) Kruger flaps,
(c) Fowler flaps.

6 **Undercarriage struts utilising compressed air and oil for springing and damping are known as:**
(a) Oleo struts,
(b) Hydraulic jacks,
(c) Pneumatic actuators.

7 **Disc brakes are preferable to drum units because:**
(a) They last longer,
(b) They do not fade to the same extent under heavy use,
(c) They can be hydraulically operated.

8 **In a light aircraft hydraulic pressure for the brakes is provided by:**
(a) Master cylinders operated by the pilot's hand or feet,
(b) An engine driven pump,
(c) A venturi tube.

9 **Control locks are provided:**
(a) To prevent unauthorised use of the aircraft,
(b) To protect the control surfaces while moving the aircraft in and out of the hangar.
(c) to avoid wind damage while the aircraft is parked on the ground.

5 Airframes

10 Some high wing monoplanes have struts running from the bottom of the fuselage to the mainplanes for the purpose of:
  (a) Providing a foothold to assist while refuelling,
  (b) Supporting the wing while the aircraft is on the ground,
  (c) Supporting the wing while the aircraft is on the ground and when it is flying.

11 Frise ailerons reduce aileron drag by:
  (a) Moving down through a larger angle than the up-going aileron,
  (b) Moving up through a larger angle than the down-going aileron,
  (c) The leading edge of the up-going aileron protruding below the wing surface and causing drag on the inside of the turn.

12 A cantilever wing supports an aircraft with the aid of:
  (a) Bracing wires,
  (b) One or more deep spars,
  (c) One or more struts.

Chapter 6
Piston Engines and Propellers

Although many of the early aeronautical thinkers envisaged man-powered flight (or even bird-powered flight), such pioneers as Sir George Cayley and, more than a century later, the Wright brothers, recognized that the key to success lay in the discovery of a suitable engine. At the time steam was both low powered and cumbersome and the gunpowder engine proved a blind alley. During the eighteenth century engineering techniques were limited so that experiments with hot-air engines proved unsuccessful, although in many ways, this type of prime mover may be regarded as the link between steam and the modern internal combustion engine, which more or less overnight made petrol, hitherto regarded as of little value, into the life-blood of industrial nations.

While it is not necessary for pilots to have a deep technical knowledge of the theory and design of the petrol engine it is nevertheless important that they should understand the basic principles of the power plant in their aircraft.

Principle

The piston engine produces power by converting fuel (petrol or diesel oil) into heat and heat into energy. The energy is harnessed by mechanical means and the resultant power is then transmitted by a rotating shaft which is linked to the driving wheels of a car or made to rotate the propeller in a boat or aeroplane. It is self-evident that when ignited a liquid fuel will burn and produce heat but how the heat is able to provide mechanical energy will be less obvious until it is realized that heat can be applied to air and heated air will expand very considerably exerting great pressures if contained in any way.

Carburation

To provide perfect combustion in the engine petrol and air must be mixed in a ratio of approximately fifteen parts of air to one part of

207

6 Piston Engines and Propellers

petrol by weight. Too much petrol causes a **Rich** mixture (which may be recognized by black exhaust smoke). This is both wasteful of fuel and damaging to the engine because heavy carbon deposits build up in the combustion areas. Conversely too little petrol in relation to air causes a **Weak** mixture. This, too, is damaging to the engine; instead of the mixture burning at an even rate on ignition an explosion occurs causing severe strain on the engine accompanied by overheating and loss of power. In most piston engines fuel is mixed with air in the **Carburettor**, a simple example being illustrated in Fig. 111.

Fuel from the aircraft tanks is pumped to the carburettor where it enters a device similar in design to a domestic w.c. cistern called a **Float Chamber**, its function being to maintain the fuel at the correct level in the **Jets**. These jets create an atomizing action rather similar to that of the nozzle of a scent spray except that, in a carburettor, fuel is drawn out of the jets by a drop in pressure created by the venturi tube shown in the illustration. It is usual to provide a **Slow Running Jet** for the purpose of engine idling and a **Main Jet** for normal operation. On some carburettors when the throttle is opened for maximum power an additional **Power Jet** is brought into operation.

A reduction in air density will occur as the aircraft gains altitude and since this would result in an over rich fuel/air mixture the carburettor is provided with a **Mixture Control**, a type of valve which can vary the fuel/air ratio. An adjustable **Throttle Butterfly** positioned between the jets and the engine is linked to the pilot's

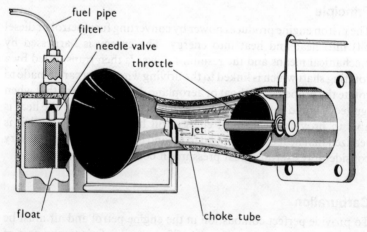

Fig. 111 Simple carburettor.

throttle control so that the amount of mixture entering the engine may be regulated, this in turn determining the power output to the propeller.

The Four-Stroke Cycle

Most engines fitted to light, as opposed to microlight or ultra-light aircraft are of the **Four-stroke** type and the meaning of the term will be explained in the following text.

When the correct mixture for various conditions of height and power setting has been produced it must be transferred to the engine for combustion. As the term implies, internal combustion means that the mixture is burned within the engine proper, as opposed to steam engines where fuel is consumed under a boiler and the resultant steam is transferred to the engine.

In most car and light aircraft engines the petrol/air mixture is inhaled by the engine which in many respects is similar in layout to a bicycle pump. Put into simple terms a **Piston** is arranged to slide up and down within a **Cylinder** which is closed at one end. There must be sufficient clearance between the walls of the cylinder and the sides or **Skirt** of the piston to allow for metal expansion as the engine reaches its working temperature. At the same time a gas-tight fit must be maintained and the piston is therefore provided with a number of **Piston Rings** for the purpose. The rings are made of a high quality cast iron, springy in nature so that they press lightly against the cylinder walls. They are split to allow for expansion.

During starting the piston is drawn down the cylinder and the resultant decrease in pressure causes the engine to suck in mixture from the carburettor through an **Inlet Valve** provided for the purpose in the **Cylinder Head**. This phase of the operating cycle is known as the **Induction Stroke**.

Early in the development of the petrol engine it was found that power could be greatly increased if the mixture charge was compressed in the cylinder prior to combustion and the next phase is called the **Compression Stroke**. The inlet valve remains closed during this stroke and the piston moves to its uppermost position compressing the mixture into a small area of the cylinder head called the **Combustion Chamber**. The degree of compression is referred to as the **Compression Ratio**, this being the ratio between the compressed volume (referred to as 1) and the volume of mixture before compression. Thus a compression ratio of 6:1 means that six volumes of mixture are compressed into one volume.

While high compression ratios yield extra power, special fuels are needed to prevent **Detonation**, i.e. explosion after the engine ignites the mixture. Detonation may be recognized as the familiar 'pinking' experienced when a car is made to pull hard in top gear while operating on low grade fuel. There is a tendency in design towards higher compression ratios both in car and aeroplane engines and while 6:1 was once considered average 8:1 or even 10:1 are now not uncommon. As a guiding principle the higher the compression ratio the higher must be the lead content of the fuel. This is expressed as an **Octane Number** and the pilot must ensure that the correct grade of fuel is put into the tanks of the aircraft since too low an octane rating will produce detonation while too high a rating may damage the cylinders and exhaust valves.

Having inhaled a charge of mixture (stroke 1) and then compressed it (stroke 2) the engine is now ready to produce power by burning the mixture when the resultant heat will expand the already compressed air.

The mixture is ignited by a **Sparking Plug** which is situated in the cylinder head. At the appropriate time a high tension current generated by the **Magneto** (or ignition coil in a car) is led to the plug causing a spark to jump across its electrodes and ignite the fuel/air mixture in much the same way as the spark igniter fitted to most domestic gas cookers. The combustion that follows is not to be confused with an explosion (a sudden release of energy) but it should be, imagined as a progressive burning of the fuel, spreading within the combustion chamber like the water ripple which results when a stone is dropped into a pond. The spreading flame heats the air, creating very high pressures within the confines of the cylinder head, and the expanding gas forces the piston down the cylinder with a great deal of energy. This third phase is known as the **Power** or **Ignition Stroke**.

The cylinder is now full of burned mixture (mainly carbon monoxide gas) which has to be removed before a fresh charge can be sucked in for another power stroke and the piston must again travel up the cylinder forcing out the hot gases through the **Exhaust Valve** which is situated in the cylinder head adjacent to the inlet valve. On completion of the fourth stroke the engine is ready to begin the cycle again. From the foregoing it will be seen that in the Four-stroke engine described four strokes occur in a complete cycle only one of which produces power.

An alternative type of engine, the **Two-stroke**, provides power every second stroke. Such engines are of very simple design with no moving valves. Having twice the number of power strokes one might imagine that, size for size, two-stroke engines would be twice as powerful as four-stroke units. However, this is not the case; two-stroke engines do not burn and exhaust the mixture efficiently. Also, lubrication is attained by mixing oil with the petrol and, at times, this can lead to fouled plugs. While there are some new, improved two-stroke engines being developed, at present most aero engines are of the four-stroke type.

It is of course necessary to convert the up-and-down or **Reciprocating** motion of the piston to a rotary movement of the propeller shaft. Additionally a mechanism must be provided to open and close the inlet and exhaust valves at the correct time and a high tension current has to be switched to the sparking plug when the compressed mixture is ready for ignition. So it is now appropriate to consider the engine in greater detail. The student will by now have noticed that most terms relating to engines are both descriptive and self-explanatory.

Provision of Rotary Motion

In a bicycle the reciprocating motion of the rider's legs is transmitted to the crank via a pair of pedals so converting an up-and-down motion into a rotation which is transmitted by a chain to the rear wheel. In a piston engine (petrol, diesel or steam) the crank and pedals are replaced by a **Crankshaft** and the 'legs' of the engine are provided by a **Connecting Rod** which links the crankshaft to the piston (the muscles of the leg). To allow for the sweeping motion of the crankshaft as it rotates within its **Crankcase** provision is made for the connecting rod to pivot on a **Gudgeon Pin** inserted across the axis of the piston. The pin passes through the **Small End** of the connecting rod and the **Crank Pin** (equivalent to the pedals on a bicycle) is free to rotate in the connecting rod **Big End** bearing (Fig. 112).

When the engine is running at 3000 RPM the piston and connecting rod must go up the cylinder, stop, descend down the cylinder and stop again (ready for the next ascent) fifty times every second and it is essential to counterbalance the crankshaft so that vibration is kept to a minimum.

Valve Mechanism

Most aero engines have **Overhead Valves**, that is to say, the valves are arranged to open and close in the top of the cylinder head, top being

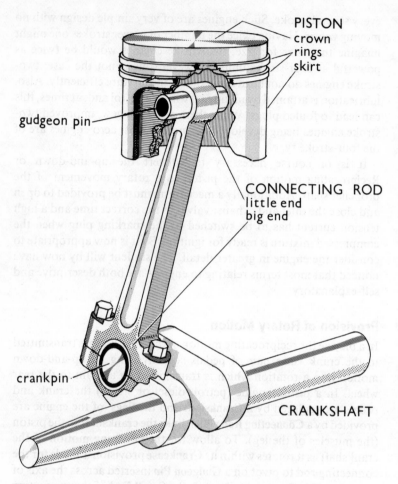

PISTON
crown
rings
skirt

gudgeon pin

CONNECTING ROD
little end
big end

crankpin

CRANKSHAFT

Fig. 112 Piston, connecting rod and crankshaft assembly.

in relation to the engine which may have its cylinders mounted upright (i.e. car fashion) or, in the case of aero engines, horizontally or even inverted. The valves, which resemble long stemmed mushrooms, are kept closed by powerful **Valve Springs**. There is a tendency for valves to bounce as they snap shut and this is overcome by having two or perhaps three coil springs fitted one inside the other over the valve stem. As each spring has a different diameter their vibration characteristics cancel one another.

To achieve a good gas-tight fit the valves are ground and polished into their **Valve Seats**. When opened the inlet valve allows mixture to flow from the carburettor through the **Inlet Port** to the combustion chamber. Similarly the exhaust valve allows exhaust gases to flow from the cylinder through the **Exhaust Port** to the exhaust system. The stems of the valves slide in **Valve Guides** (Fig. 113).

It is essential that the valves open and close with a high degree of precision relative to the position of the piston and logically enough

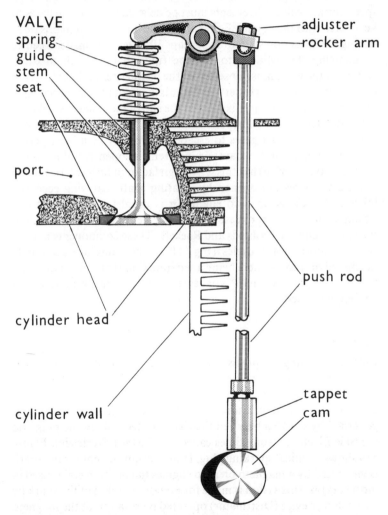

Fig. 113 Valve and its associated operating mechanism.

this is called **Timing**. When the crankshaft is in its uppermost position with the piston at the top of its travel the assembly is said to be at **Top Dead Centre** while the other extreme (piston at the bottom of its stroke) is called **Bottom Dead Centre**. Valve openings and closings are timed through suitable gearing to occur at a particular number of degrees before or after top or bottom dead centre according to the design of the engine. There is no need for the pilot to understand fully the complex design factors involved other than to realize that fuel/air mixture (or any gas) has inertia and with an engine running at high speed certain valve movements must be delayed to allow the gases to keep pace with the engine. For example the inlet valve remains open during the induction stroke for perhaps 10° after bottom dead centre so allowing the cylinder to become fully charged with mixture. Naturally the valve must close before the piston rises in the cylinder to any appreciable extent otherwise the mixture would be pumped back through the carburettor. Likewise at the end of the exhaust stroke the exhaust valve will remain open for say 10–15° after top dead centre to allow all the exhaust gases to escape. There may be a brief period when both inlet and exhaust valves remain open simultaneously (called **Valve Overlap**) but this is a design factor of little interest to the pilot. The valves are opened by rotating shafts carrying eccentric lobes called **Cams**. In aero engines it is usual for the cams to be situated within the crankcase, their movement being transmitted to the valve stems via **Pushrods** and **Rockers**. These components may be seen on most motor cycles (Fig. 113). The cams are accurately machined on a shaft, there being a separate **Camshaft** for exhaust and inlet valves. The camshafts are geared to the crankshaft and run at half engine speed.

Ignition

Like valve timing the precise moment when the mixture is ignited is all-important. The spark has to be of sufficient intensity to ignite the petrol vapour and the sparking plug must provide the spark under conditions of high pressure and great heat. Length of spark is governed by the **Gap** between the **Central Electrode** of the plug and the **Side Electrode** (sometimes called the **Earthed Electrode**). Figure 114 shows a typical sparking plug. High tension current for the spark is generated by a magneto on aero engines (an ignition coil is used in most cars) and provision is made for precise emission of the spark by a switching device (**Distributor**) operated by rotation of the magneto shaft. The device incorporates a pair of platinum **Points** which is

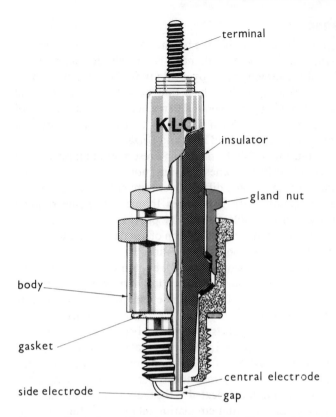

Fig. 114 Typical sparking plug, part sectioned to reveal its construction.

accurately adjusted to ensure that the spark occurs at the right time. It would be expected that ignition is timed to occur just after the piston has compressed the mixture and started to descend but it should be remembered that combustion in a piston engine is a progressive burning of the fuel which commences gradually, building up as the flame spreads across the combustion chamber. Time is therefore required for full combustion to develop so that the spark is arranged to occur 10–15° before top dead centre (BTDC). The higher the operating **RPM** the more must the spark be **Advanced** (in cars it is the practice to fit an automatic advance/retard control which caters for various engine speeds and load conditions).

Lubrication

It is not generally realized that the balance between smooth running of an engine and irreparable damage is entirely dependent upon separation of the moving parts by a thin film of oil, often no more than one-thousandth of an inch in thickness. Dry metal-to-metal contact of fast moving parts is bound to result in the rapid build-up of heat through friction, to such an extent that within a matter of seconds the components would melt and **Seize**.

Provision must therefore be made for a reservoir of lubricating oil of the correct **Viscosity** (thickness). It must not be so thick when cold that engine starting is made difficult yet it must be able to provide adequate lubrication when it becomes thinned at engine running temperatures. An oil cooler is usually provided to prevent over-thinning of the oil. It may take the form of a separate oil radiator or the oil tank itself can be cooled by the airflow.

The oil is carried in a **Sump** (reservoir) built into the bottom of the crankcase and an engine-driven pump is provided to circulate oil under pressure through drillings in the engine which communicate with the main bearings, big-end bearings, cam-shaft bearings, valve rockers and other moving parts. Badly worn bearings allow the oil to seep out and cause a drop in oil pressure so that the oil pressure gauge is often a good indication of their condition. In addition to the pressure lubrication already mentioned certain parts such as the cylinder walls are **Splash Lubricated** as the crankshaft rotates and the big ends dip into the oil. After circulation oil may be returned to the sump by draining or **Scavenging**. Some engines have a separate **Scavenge Pump** for the purpose. While performing its function oil collects minute particles of metal and carbon deposits so that **Filters** must be provided to keep it clean. These can be of the felt-element type which is replaced at regular intervals or there may be a permanent filter element of thin, interleaved metal discs which is cleaned by rotating a T-shaped handle on the filter casing.

Summing up lubrication, the oil has these vital tasks to perform in the engine:

1. Provision of a protective film which 'plates' itself on moving parts so preventing metal-to-metal contact.
2. Dissipating heat generated by the moving parts.
3. Removal of impurities such as metal particles, carbon or other deposits which result from engine operation. Modern oils have a detergent action.
4. Provision of a gas-tight seal between valves and valve guides and

piston rings and cylinder walls. To prevent oil working past the pistons and entering the combustion chamber a special **Oil Control** ring is sometimes fitted around the lower part of the piston skirt.

Cooling

Quite apart from the heat generated by fast moving parts is the very considerable heat resulting from combustion of the mixture. When running at 3000 RPM an engine such as the single-cylinder unit under discussion would ignite every two revolutions or

$$\frac{1500}{60} = 25 \text{ times per second.}$$

When there is a need to burn more fuel (e.g. to provide extra power for the climb) engine temperature increases and cooling becomes even more essential. Unless the surplus heat generated were dissipated in some way it would only be a matter of minutes before the cylinder head became red hot and the engine seized.

In a car the problem is solved by surrounding the cylinder and cylinder head with a series of passages cast in the cylinder block. Water pumped through these passages or **Water Jacket** is returned through a radiator where the heat is dispersed by the air flowing through it. While **Liquid Cooling** has in the past been adopted for some very successful high performance aero engines (notably the Rolls-Royce Merlin) large engines are now practically confined to the gas turbine family. Lower powered aero engines have almost without exception relied upon a cooling system which dispenses with the radiator and its tank, the water jacket around the cylinder and the coolant itself. This simpler and lighter method, in fact the one used on motor cycles and small petrol mowers, is called **Air Cooling**. Fins are machined around the outside of the cylinder barrel and cylinder head and a system of sheet-metal **Baffles** guides the airflow through the engine cowling and around the fins carrying away the excess heat through the rear of the engine bay. Unfortunately some two-thirds of the fuel energy in an internal combustion engine is wasted, only the remaining third being converted to power.

All four-stroke engines, large or small, are based upon the simple unit described in the foregoing paragraphs. How these principles are applied to the specialized requirements of the aero engine is explained in the next section of this chapter, but before reading on the student is advised to study and understand fully the four-stroke cycle which is illustrated on page 219 (Fig. 115).

The Aeroplane Engine

So far the engine has been considered as a single-cylinder unit, producing, in the four-stroke design, one power stroke every two revolutions. Two cylinders arranged with a double crank so that one piston is at TDC (top dead centre) while the other is at the bottom of its stroke will give a power stroke every revolution and a four-cylinder engine fires every half revolution. Power output from an engine is dependent upon the amount of force exerted by each piston during the power stroke multiplied by the number of times per minute the force is produced.

By definition time is an important factor of horsepower. Remember one horsepower is the amount of power required to raise 33,000 lb through a height of one foot *in one minute*. Equally one horsepower is the equivalent of raising 66,000 lb through six inches in one minute or 330 lb through 50 ft in half a minute. In each case the amount of power is the same: 33,000 ft/lb/min. It therefore follows that the engine designer may obtain the required power from his engine by:

(a) using a number of small pistons and cylinders and running the engine at high speed;
(b) using fewer but larger pistons and cylinders at the same speed;
(c) reducing the speed of the engine and increasing the size of the pistons and cylinders.

Of course other measures can be taken to increase the power of an engine but these are often at the expense of long life and reliability, both essential requirements in an aero engine. While a six-cylinder engine is likely to be smoother running than a four-cylinder unit of the same power, aero engines developing 180 HP or less are almost exclusively designed with four relatively large cylinders, such engines being more robust than would be the case with six smaller piston/connecting rod assemblies.

The aeroplane engine differs from its automobile counterpart in a number of respects. Whereas the car engine is expected to operate under conditions of constantly changing speed and load, the aero engine has to run for long periods at a steady cruising power. High RPM, desirable in a car engine, are avoided in an aero engine because

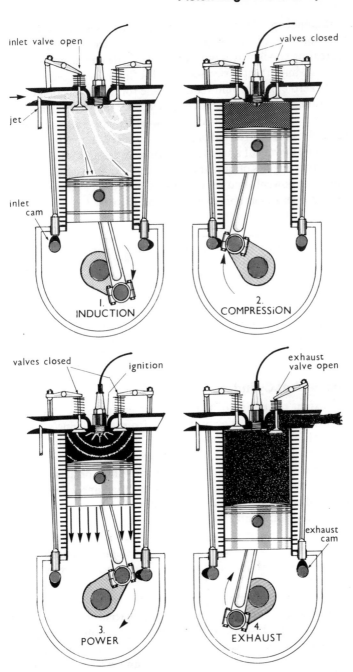

Fig. 115 The four-stroke cycle.

when rotational speeds exceed certain limits the tips or possibly a greater proportion of the propeller blades will exceed the speed of sound, creating shock waves which both dissipate engine power and reduce the all-important thrust required for flight. In consequence a high-RPM aero engine would require some form of **Reduction Gear** between crank and propeller shaft but such gears tend to be damaged by harmonic vibration that occurs between engine and propeller. Consequently, the requirement of power at low RPM is met by having a large engine. By way of illustration it is interesting to compare aero engines and car engines of similar power output. In each case it will be seen that while power for power the aero engine is twice the size of the car engine it operates at approximately half the speed.

	Austin Rover MG EFi car engine	Lycoming 0–235–LZC aero engine	Ford Granada 2.8 Ghia X car engine	Lycoming IO–360–C1C6 aero engine
BHP	115	112	150	160
RPM	5800	2600	5500	2700
Capacity (cc)	1994	3823	2792	5250
No. of cylinders	4	4	6	4

Arrangement of Cylinders

To some extent the design of the aeroplane engine is dictated by a need to arrange Thrust (i.e. propulsion from the engine(s)) in a suitable position relative to Drag (the total resistance of the wings, fuselage, etc. during flight) and although early engines had the cylinders fitted above the crankcase motor-car fashion it was soon realized that by inverting the engine a better thrust line would result accompanied by an improvement in the pilot's forward view over the nose of the aircraft. Until 1960 most of the lower-powered engines produced in Britain were of the **Inverted In-line** design, i.e. cylinders arranged in line below the crankcase. A classic example of this type of engine is the Gipsy Major in its various forms and the six cylinder Gipsy Queen.

American engines in the low-power category have been developed on rather different lines with the cylinders placed horizontally on both sides of the crankcase. One of the advantages of the **Horizontally Opposed** engine is that it is shorter than an in-line unit of the same number of cylinders, the shorter crankshaft and crankcase resulting in an appreciable saving in weight (Fig. 116). As a measure of this

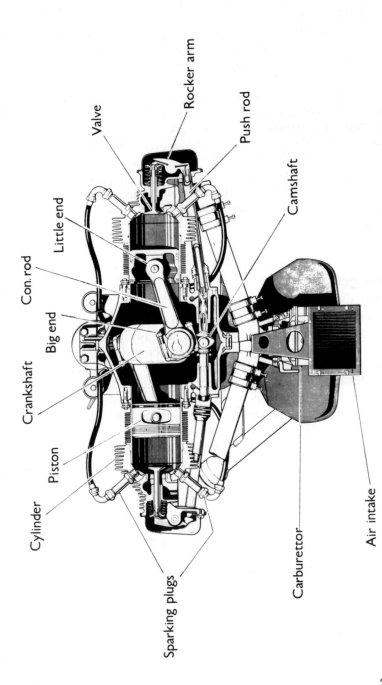

Valve

Rocker arm

Push rod

Little end

Con. rod

Camshaft

Big end

Crankshaft

Piston

Cylinder

Air intake

Sparking plugs

Carburettor

Fig. 116 Continental 0-200A horizontally opposed engine. The front pair of cylinders are shown in section.

aspect of efficiency an engine is said to weigh, for example, 1.98 lb/HP. Although horizontally opposed engines originated as low-powered units there now exists a very complete range starting from 35 HP and extending to eight-cylinder power plants (four each side of the crankcase) developing 400 HP.

High-powered piston engines are now confined to the **Radial** design where the cylinders are spaced around a circular crankcase and the connecting rods are attached to a **Master Rod**, its bearing running on a single crank within the crankcase. For power outputs in excess of 1500 HP a second row of cylinders is placed behind the spaces between cylinders in the first row when the engine is designated a **Two Row Radial**. Large radial engines are these days mainly used in Ag-planes. These various cylinder arrangements together with others no longer in current use are illustrated in Fig. 117.

Dual Ignition

With few exceptions even the most expensive motor cars run on one set of sparking plugs and a single ignition coil. For reasons of safety all aeroplane engines are required to have two separate ignition systems, i.e. two magnetos feeding two sets of sparking plugs through separate ignition leads. There is, however a less obvious reason for **Dual Ignition**. It will be remembered that the aeroplane engine is designed to develop its power at low RPM. Because of this it is invariably a large engine relative to power output so that large pistons and cylinders are used in conjunction with a combustion chamber of considerable volume. While the large volume of compressed mixture could be ignited by a single sparking plug, better combustion and therefore more power results from having two sparking plugs positioned one either side of the cylinder head. It is part of a pilot's pre-flight checks to **Run up** the engine, testing each ignition system in turn, and noting the decrease in RPM or **Mag Drop** as each switch is turned off. The maximum permitted drop in RPM per magneto, and the maximum difference in mag drop that may be allowed between magnetos will be listed in the aircraft manual. A serious drop in RPM could be caused by a defective magneto, a faulty plug lead or an unserviceable sparking plug.

When the switches are both in the ON position an increase in RPM will be seen on the engine-speed indicator thus demonstrating that dual ignition increases power on a large slow-running engine.

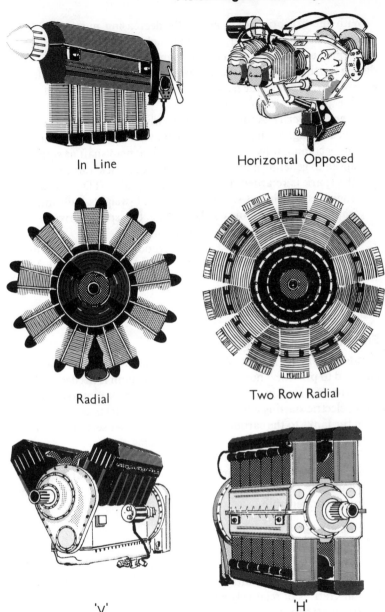

In Line

Horizontal Opposed

Radial

Two Row Radial

'V'

'H'

Fig. 117 Cylinder arrangements. First four engines are air cooled but
only the horizontally opposed example (top right) remains in
widespread use. The two large, liquid cooled engines (bottom)
and the bigger radials have given way to gas turbines.

With the exception of a gradually decreasing number of low-powered engines installed in simple light aeroplanes starting is performed electrically. Even so the engine is turned over very slowly during starting and one of the following methods has to be adopted to ensure that, although the magneto is rotating slowly, a high-intensity spark is provided for starting:

1. *Impulse Starter*. A form of magneto which is driven by the engine via a spring. As the engine is turned over for starting, the spring, on becoming fully wound, flicks the magneto shaft at sufficient speed to generate a spark. To prevent 'kick back' during starting the spark is retarded, ignition occurring after TDC. When the engine is running at some 700 RPM bob-weights built into the impulse device fly outwards locking the spring drive in the correct position for advanced ignition.

2. *High-intensity Magneto*. The method favourved by most American manufacturers is to provide magnetos of large capacity which are capable of generating a spark at low rotational speeds.

Engine-driven Ancillaries

In addition to its main task of driving the propeller an engine is required to provide a number of services. Vacuum is needed to drive the gyro-operated instruments, electric power is required for the radio, electric starting and various lights. Also, fuel must be pumped from the tanks to the carburettor. These and other services are more fully described in Chapter 7, Aircraft Systems.

The various pumps, alternators/generators, etc., are driven by the engine via a **Wheelcase**, a gearbox that is usually located at the rear of the crankcase. Some engines use a belt and pulleys to drive the alternator or generator.

Electric starter

Most light aero engines use a car-type electric starter which engages with a large, toothed ring attached to the crankshaft at the front of the engine. It is powered by an aircraft-type battery which, in the case of light aircraft, is of relatively small capacity.

To conserve the battery while on the ground some light aircraft are fitted with a socket that accepts a plug from a large capacity **Trolley-acc**, a set of batteries on wheels, which can maintain various electric services for the purpose of testing and supply power for starting. Generally, modern light aircraft have good starting systems with adequate reserves of power notwithstanding the small size of their battery.

Engine Handling

Dependent upon the complexity of the aircraft the engine is handled through some or all of the following controls:

1. Ignition switches
2. Fuel cock(s)
3. Throttle
4. Primer
5. Mixture control
6. Idle cut-off
7. Carburettor heat control
8. Electric fuel pump
9. Electric starter
10. Battery master switch
11. Propeller pitch control
12. Cooling flap control.

The correct procedures to be adopted while starting, during various flight conditions and when stopping the engine are explained in the flying training manuals which form part of this series (Volume 1). A detailed description related to the aircraft type will be found in Section 7 of the Owner's/Flight/Operating Manual.

1. *Ignition Switches.* These may take the form of separate tumbler switches similar to the usual domestic light switch with the exception that in the down position the ignition is off, the switches being UP for CONTACT. A separate switch is provided for each magneto and when hand starting only the switch controlling the impulse magneto (when fitted) should be used (see page 224). More usually, certainly in the case of light, single-engine aircraft, the ignition may be controlled through a key-operated switch which has the following positions:

OFF
MAG 1
MAG 2
MAGS 1 and 2.

With some installations the starter is engaged by turning the ignition key beyond the MAGS 1 and 2 position where it is held against spring pressure until the engine fires.

Whatever the type of switch when the ignition is in the OFF position the magnetos are earthed to the engine and so are unable to cause a spark.

2. *Fuel Cock*. Most single-engine trainers have two fuel tanks controlled by a simple ON/OFF cock. Alternatively a single multi-position cock will handle fuel from whatever tank is selected by the pilot.

3. *Throttle*. This may be of the 'plunger' or 'lever' type. An adjustable **Throttle Friction** device is often provided so that there is free movement of the throttle during taxying and sufficient friction to hold the throttle in the position set by the pilot during cruising or other flight conditions.

    In multi-engined aircraft a separate throttle is provided for each power unit.

4. *Primer*. When starting a cold engine a very rich mixture is required. The **Primer** is a small plunger-type pump which the pilot operates to inject petrol to the engine prior to cold starting. Care should be taken to ensure that the plunger is screwed home after use otherwise the mixture will be disturbed and rough running of the engine will result.

    Most light aero engines of American design may also be primed by pumping the throttle several times before starting. Rich mixture is in this case provided by an accelerator pump built into the carburettor for the main purpose of improving engine response to throttle opening.

5. *Mixture Control*. It was explained (page 208) that provision must be made to compensate for changes in air density which occur with changes in altitude. Without a mixture control to vary the fuel/air ratio the engine would run 'rich' as the aircraft climbed into less dense air. This is a factor of some importance at heights in excess of 5000 ft, materially affecting the economical operation of the aircraft. Provided the engine is running at not more than 75 per cent power the mixture may be leaned at any height to provide fuel economy. The mixture control is moved towards the WEAK position until there is a decrease in RPM when it must be moved slightly in the opposite direction until the original engine speed is restored. A **Fuel/Air Ratio Meter** may be included with the engine instruments to assist in setting optimum mixture.

6. *Idle Cut-off.* Although it is possible to stop an engine by switching off the ignition, because of their design aeroplane engines have a tendency to 'fire on' intermittently unless the throttle is opened wide as the propeller ceases to rotate. A cleaner stop will result when petrol is prevented from entering the slow-running jet and most aircraft have an idle cut-off for this purpose. It is brought into operation by moving the mixture control to the fully WEAK position.

   Before operating the idle cut-out the engine should be allowed to run at 800–1000 RPM for a few moments, so enabling it to cool slowly to an even temperature. When the engine has been stopped the ignition switches *must* be put in the OFF position.

7. *Carburettor Heat Control.* Basically the carburettor is a device for atomizing fuel and mixing it with air in the correct ratio. This function is accompanied by a marked decrease in temperature within the area of the jets, due in part to the vaporization of petrol and also to the decrease in pressure which occurs within the choke tube which, in turn, causes a drop in temperature (conversely while inflating a cycle tyre the pump becomes hot due to the rise in air pressure). When the outside air temperature is between +30°C and —18°C the cooling effect of the carburettor is sufficient to freeze the moisture content of humid air. Ice deposits may then form on the butterfly valve and surrounding areas of the choke tube. The engine will slow down and eventually stop following a **Rich Mixture Cut.**

   To overcome this hazard ducting is arranged to convey air to the carburettor from two sources.

(a) Cold air from outside the aircraft.
(b) Hot air from a heat exchanger built around the engine exhaust pipe. A two-way trap valve conveys hot or cold air to the carburettor according to the position of the carburettor heat control. (Fig. 118) The student may wonder why it is necessary to provide this control when the cure for carburettor icing is simply a matter of drawing in warm air. While some light aircraft of older designs do in fact operate in warm air during cruising flight a linkage is nevertheless arranged so that on selecting take-off power the trap valve moves to the cold air position. The piston engine relies for its power on the basic principle of expanding a mass of cold air by heating. It therefore follows that the more air is packed into the cylinders for expansion the greater will be the power developed by the engine. To a considerable extent this

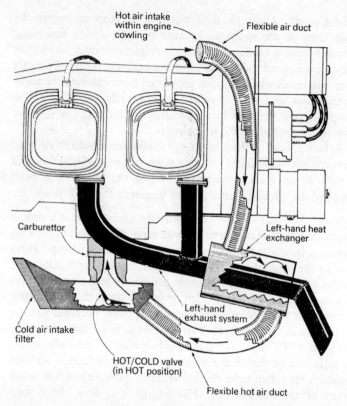

Fig. 118 Typical carburettor heat system. A similar heat exchanger is
located on the other side of the engine to provide cabin heat.

aspect of the piston engine (known as **Volumetric Efficiency**) is
dictated by the density of the air entering the cylinder and since
by pre-heating the air its density is reduced so a decrease in power
will result when hot air is selected on the carburettor heat
control. This drop in power may be demonstrated convincingly
by selecting hot air when the engine is running at or above
cruising power. Hot air selection causes a decrease in RPM.

8. *Electric Fuel Pump.* When the fuel tanks are below the level of the
carburettor, as would be the case when fuel is carried in the
fuselage below the baggage area or in the wings of a low-wing
aircraft, an extra pump must be included in the fuel system.
Although engines are fitted with a diaphragm-type pump worked
mechanically from the camshaft it is usual to have an additional

electric pump, either immersed within the tank or similar in type to the usual car pump. As a safety precaution against failure of the engine-driven unit the electric pump must be switched on during take-off and landing, or during any other phase of flight when engine failure may have resulted from malfunction of the engine-driven pump.

The engine instruments usually include a **Fuel Pressure Gauge**.

9. *Electric Starter*. Before the first engine start of the day the ignition switches are checked in the OFF position, then the engine should be turned over by hand:

(*a*) to free the engine from the effects of oil film adhesion and so reduce the load on the starter, and

(*b*) to check that oil has not collected in the cylinders since severe damage would result should the engine start and one or more of the pistons come up against oil. **Hydraulicing**, as the condition is known, can remove the affected cylinder head.

The starter should be operated for brief periods only, allowing time for battery recovery after each attempt to start the engine.

10. *Battery Master Switch*. This may take the form of a simple **Master Switch** which brings the battery into and out of circuit or more usually on a modern aircraft, it will be of the **Split Rocker** type, the left-hand side controlling the alternator, the right half dealing with the battery. In normal use both halves of the switch are operated together (see page 255, Chapter 7, Aircraft Systems).

11. *Propeller Pitch Control*. Although most training aircraft are fitted with fixed pitch propellers some light touring aeroplanes have **Constant Speed Propellers**. These are described on page 238.

12. *Cooling Flap Control*. Usually only found on aircraft with engines of more than 200 HP the cooling flaps control the flow of air through the engine cowling so enabling the pilot to maintain the engine at its best working temperature.

Fuel Injection

A more refined way of dispensing fuel to an engine than by using a carburettor is **Fuel Injection**. Whereas fuel/air mixture is drawn out of the carburettor venturi by the engine, causing a drop in temperature and a risk of carburettor ice due to decreased pressure

6 Piston Engines and Propellers

and the evaporation of fuel (page 227) fuel injection systems work on a rather different principle.

An engine-driven pump, backed up by an electric fuel pump which is also used for priming the engine, supplies accurately metered fuel under pressure to a **Fuel/Air Control Unit**. There it is mixed with the correct amount of air for onward distribution to the engine via injection nozzles located within the inlet valve chamber of each cylinder.

The advantages of fuel injection are:

(a) Better fuel economy than carburettor engines of the same type;
(b) Freedom from induction icing.

There are, however, some disadvantages. Fuel injected engines are more expensive than those with carburettors. Also they tend to idle more roughly and can be difficult to start when hot unless the individual engine and its characteristics in this respect are well known by the pilot.

Fuel injected engines are provided with a dual **Manifold Pressure/Fuel Flow Indicator** which is two instruments within one case. The fuel flow part of the instrument indicates fuel flow in gallons per hour (usually US gallons).

Starting Fuel Injected Engines

Procedures differ slightly from one engine type to another but a typical method of starting when the engine is cold would be:

1. Brakes ON, switches CONTACT, throttle CLOSED, mixture RICH.
2. Electric fuel pump ON.
3. Open throttle until 6 gallons/hr (or as recommended in the aircraft manual) is indicated on the fuel flow meter. Switch off the fuel pump.
4. Close the throttle and set slightly open for starting.
5. Set the mixture to Idle Cut-off.
6. Operate starter. When engine fires move mixture control to RICH. Then adjust the throttle to produce a steady idle.

Turbocharging

The engines so far described derive their fuel/air mixture by drawing in a charge as the pistons descend during the induction stroke. Such engines are known as **Normally Aspirated**.

230

As an aircraft climbs and the air density decreases so there is a corresponding reduction of power until an altitude is reached where insufficient power is available to continue the climb. When an aircraft has reached an altitude where the rate of climb has fallen below 100 ft/min. this is known as its **Service Ceiling** (as opposed to **Absolute Ceiling** which is the altitude at which the aircraft ceases to climb).

Typical figures for a small engine would be:

sea level	110 HP
8 000 ft	75 HP
12,000 ft	62 HP

Although the student pilot will learn to fly in aircraft powered by simple, normally aspirated engines, after gaining a PPL the opportunity may arise to pilot a light tourer fitted with a **Turbocharged** engine. Turbocharging was originally developed to improve the high altitude performance of aircraft but like so many technical developments of this kind (disc brakes being another example) the fruits of aeronautical knowledge have been applied to cars. Turbochargers are fitted to improve the performance of cars, mainly in terms of acceleration and top speed. In aircraft their purpose is:

1. To improve take-off performance, particularly while operating from 'hot-and-high' airfields where reduced air density can seriously lengthen the take-off distance.
2. To compensate for decreased air density with height so that:
    (*a*) Rate of climb can be maintained, and
    (*b*) Maximum cruising power is available at altitudes where there is a large gain in TAS for any IAS.

To make full use of item 2*b* would entail cruising at altitudes of 20,000 ft or more and at such levels oxygen masks would have to be worn. However, the ability to maintain maximum power, typically up to 15,000 ft does provide the aircraft with a non-reducing rate of climb up to the altitude where maximum climb power can be maintained as well as high cruising speeds and good field performance under adverse conditions.

Risk of Over-boosting

When an aircraft with a normally aspirated engine is taking off from an airfield located near sea level, on a day when standard atmospheric conditions exist, the maximum manifold pressure (i.e. with the throttle fully open) will be approximately 29.9 in.

A turbocharged engine utilizes exhaust gases to drive a turbine which is directly coupled to a small supercharger. The supercharger supplies fuel/air mixture under pressure to the engine.

To provide sufficient pressure so that maximum engine power can be maintained at, say, 15,000 ft or more, excess pressure is available at lower levels, particularly during take-off when often the aircraft will be within a few hundred feet of sea level. The manifold pressure gauge will be 'red lined' at 32–39 in. (according to engine type) and while more complex turbocharger systems are fitted with automatic limiters which prevent over-boosting the more simple installations likely to be encountered in light touring aircraft are not. In these a warning light is fitted and there is a red line on the manifold pressure gauge. Care must be exercised not to exceed the maximum pressure at any time since this can seriously affect the life of the engine.

Running-down Procedure

The turbocharger is lubricated by the engine oil system, consequently after shut-down, the turbine/supercharger will continue rotating at high speed for several minutes without a supply of pressure oil.

To avoid possible damage during the shut-down it is essential to allow the engine to idle for several minutes at 1000 RPM or so, thus ensuring that the turbocharger is rotating at the lowest possible RPM before it is deprived of pressure oil.

This very brief description of turbochargers is intended as no more than an introduction to the subject since it does not concern the student pilot. However, the student pilot of today is the private pilot of tomorrow and there is some very advanced equipment available within light/general aviation. Fuel injection, constant speed propellers and turbocharging are all commonplace. A typical turbocharged engine is illustrated in Fig. 119.

Propellers

Light trainers are fitted with simple, **Fixed Pitch** propellers which are usually of metal although some wooden examples are in use. However, because they are intended to train sailplane pilots many of the motorgliders have three position propellers which can be adjusted in flight to provide:

Fig. 119 Lycoming TSIO-540 six-cylinder engine developing 350 hp. The turbocharger is mounted high on the right and at the front of the engine (left) may be seen the gear ring for the electric starter. (Courtesy: C.S.E. Aviation Ltd., Oxford Airport).

Fine Pitch	for the take-off, climb and landing.
Coarse Pitch	for cruising.
Feather	for use when the engine has been stopped to demonstrate gliding and the use of thermals.

The great majority of student pilots training for a PPL will learn to fly on a conventional trainer with a fixed pitch propeller but whatever the type it should be understood that propellers are potentially lethal and they must be treated with respect at all times. The golden rules are:

1. NEVER walk within the plane of rotation, even when the engine is stopped.
2. NEVER allow passengers or others to approach the cabin from in front of the wing when the engine is running.
3. NEVER assume the ignition is off. The propeller must not be turned over by hand until the ignition is carefully checked to be OFF and the mixture is in Idle Cut-off.

Propellers are also vulnerable to damage and the following section explains how to avoid this.

Propeller Care

Like any mechanical device that is made to rotate at high speed a propeller is sensitive to balance. Before it is fitted to the propeller shaft the propeller will have been accurately balanced to ensure low levels of vibration. Propellers are vulnerable to stone damage. Also some aircraft provide less than adequate tip clearance while taxying and undulating ground or the transition from grass to tarmac can, if care is not exercised, cause a blade tip to strike the ground.

Blade tip damage, or impact from stones, can seriously affect the balance of the propeller; if, during the run-up and engine checks prior to take-off unusual vibration is experienced the flight must be abandoned and the aircraft taxied back for the attention of an engineer.

To safeguard the propeller pilots should take these precautions:

Engine checks. Never attempt to run up the engine while the aircraft is parked on an area of loose stones. These tend to be lifted into the propeller disc during power checks and serious damage to the blades can result.

Taxying. For the reason previously given avoid, whenever possible, taxying over ground that has loose stones which may be thrown up by the wheels or lifted by the propeller itself.

When making the transition from grass to hard surface or *vice versa* there is often a difference in levels. Cross the edge slowly and at an angle of 40–60°, thus ensuring that one wheel at a time makes the transition from surface to surface. This will help avoid risk of pitching the nose down and striking the propeller tips on the ground.

Handling. Always treat a propeller as LIVE. Check, and double check that the ignition is OFF and the mixture is in Idle Cut-off before any attempt is made to turn the propeller by hand.

Propeller theory

To fly the aeroplane power from the engine must be converted into thrust. A jet provides thrust by accelerating a relatively small diameter of air to high velocity in the opposite direction to flight. The propeller, on the other hand, is rotated by the engine to produce a large diameter air mass which is moved at relatively low speed.

To the layman with only a passing interest in aviation it is sufficient to say that the propeller screws itself through the air rather like a screw being driven into wood but the pilot under training should know rather more about propellers than that.

A study of propeller aerodynamics is complicated by the number of variables involved. Nevertheless, if these variables are understood, each in isolation, the complete picture will soon emerge.

Blade Path

A little thought will reveal that at any particular engine RPM each section of the blade will be moving at a different speed, being slowest near the propeller shaft and fastest at the tips. At, for example, 2500 RPM a blade section 12 in. from propeller centre would be moving at 155 kt while at the tips, located 3 ft from centre in the case of a 6 ft diameter propeller, the speed would be 465 kt.

Propeller blades are of airfoil section and they are subjected to forces similar in nature to those of a wing. Unlike the wing, which in straight flight experiences a constant airspeed from tip to tip, for reasons previously explained a propeller blade is subjected to airspeeds that increase from blade root to blade tip. To compensate for this blade angle must vary accordingly and if a propeller is studied it will be seen that blade angle relative to the plane of rotation is at its maximum adjacent to the propeller **Hub** or **Boss**, gradually reducing

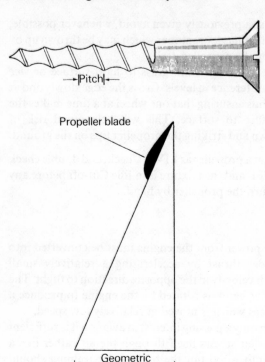

Fig. 120 Geometric Pitch may be likened to the pitch of a wood screw.
In this illustration the propeller and the screw rotate in opposite
directions while moving forward.

towards the tips. Blade angle measured in relation to the plane of
rotation is known as **Geometric Pitch**. It may be likened to the pitch of
the threads machined on a wood screw (Fig. 120).

Blade Pitch

Like any airfoil a propeller blade can only give its best performance
over a relatively narrow range of angles. In the case of a wing best
L/D ratio is usually attained at an angle of attack in the region of 4°.
Increase the angle and although lift becomes greater, drag increases
at a faster rate.

In propeller terms lift becomes thrust and drag is known as **Torque
Reaction**, i.e. the propeller's opposition to the twisting power of the
engine. To a considerable extent the pilot may control the angle of
attack of the wings by flying at speeds and under conditions

corresponding to best L/D ratio (remember that at any particular weight the aircraft has a speed that corresponds to a specific angle of attack). Propeller blade angle is more complex because it is affected by the forward speed of the aircraft.

The Propeller Helix

When the engine is running at, say, 2000 RPM and the aircraft is standing still with the parking brake on, blade angle will be stalled over much of its length. Each airfoil section of the blade will describe a circular path according to its position from root to tip.

To simplify this explanation imagine an airfoil located 2 ft from the centre of the propeller. The brakes are released and as the aircraft moves forward the flight path of the blade airfoil changes from a circle to a helix, a large spring, coils tightly packed at first but progressively stretching as the speed increases. As the helix becomes more stretched with speed so the angle of attack of our specimen blade section reduces until it moves towards the ideal 4–5°. Naturally, as the blade angle decreases so does its drag and the reduction in torque reaction makes itself present in the form of an increase in engine RPM. Although full throttle will have been applied for the take-off, in the initial stages, when parts of the blades are stalled, there is so much torque reaction (blade drag) that the engine is incapable of developing its maximum RPM.

Assuming the throttle is left alone and the speed is allowed to increase beyond normal cruising, possibly in a shallow dive, the blade helix will become further stretched and blade angle could reduce to the point where the airflow is windmilling the propeller and allowing the engine to exceed its maximum RPM. The foregoing sequence is illustrated in Fig. 121.

Although in the case of fixed pitch propellers, such as those fitted to most trainers, blade angle relative to the disc of rotation (i.e. geometric pitch) remains unchanged whatever the speed of the aircraft or engine RPM at the time, blade angle of attack alters very radically according to the helical path along which the blade 'flies'. When the blade helix is compressed (as it would be in the early stages of take-off) blade angle of attack will be large. At high airspeeds the helical path of the propeller blade will be stretched, the helix angle will be greater and, since geometric pitch has remained unchanged, propeller blade angle of attack will be less.

The following figures will illustrate the foregoing. They need not be remembered by the student pilot, their purpose being to explain the variables of propeller aerodynamics and to explain why a fixed-pitch

Airscrew (the more correct name for a propeller) can only be efficient under one set of conditions. The figures quoted in the table relate to Fig. 122 which, in the interests of clarity, is not drawn to scale.

Speed	Forward dist. per prop rev.	Dist. along helix per prop rev.	Helix angle	Blade angle of attack
Blade airfoil 2 ft from propeller centre				
Rotating at 2200 RPM				
Brakes On	Nil	12.57 ft	Nil	24.1°
40 kt	1.842 ft	12.7 ft	8.34°	15.66°
60 kt	2.763 ft	12.87 ft	12.39°	11.61°
80 kt	3.684 ft	13.09 ft	16.33°	7.67°
100 kt	4.6 ft	13.39 ft	20.1°	4.0°

It will be noted that in the early stages of the take-off blade angle is such that many parts of the propeller are stalled. Furthermore, because changes in airspeed alter blade angle of attack and therefore torque reaction (blade drag) it is necessary to adjust the throttle to obtain the required engine RPM whenever there has been a change in airspeed (e.g. during a gentle descent). To overcome these short-comings and to provide an optimum blade angle of attack under most flight conditions some aircraft are fitted with a **Constant Speed propeller**.

Constant Speed Propellers (*This section is not part of the PPL Syllabus.*)

The constant speed propeller does more than allow the pilot to alter its blade pitch. A **Constant Speed Unit** ensures that provided enough engine power has been set to turn the propeller at a minimum of, usually, 1500 RPM, the engine will remain at the selected RPM irrespective of airspeed or throttle setting.

Because throttle no longer adjusts engine RPM an additional instrument is provided for that engine control. Power setting in aircraft fitted with a constant speed propeller entails adjusting two values:

Propeller RPM: these are set on the **Propeller Lever** with reference to the usual RPM indicator.

Manifold Pressure: the pressure that exists in the induction manifold

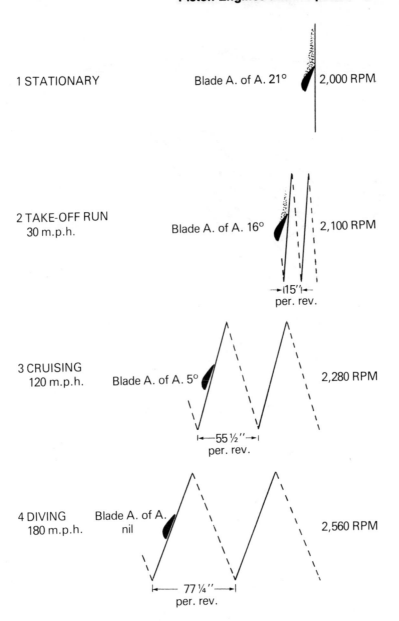

1 STATIONARY   Blade A. of A. 21°   2,000 RPM

2 TAKE-OFF RUN
  30 m.p.h.   Blade A. of A. 16°   2,100 RPM
                                ←15"→
                                per. rev.

3 CRUISING
  120 m.p.h.   Blade A. of A. 5°   2,280 RPM
                                ←55½"→
                                per. rev.

4 DIVING   Blade A. of A.
  180 m.p.h.   nil   2,560 RPM
                                ←77¼"→
                                per. rev.

Fig. 121 The propeller blade and its helix. Solid lines denote path of
          blade near the observer. Broken lines are on the opposite side
          of the helix. Note the partially stalled blade in the first two
          diagrams.

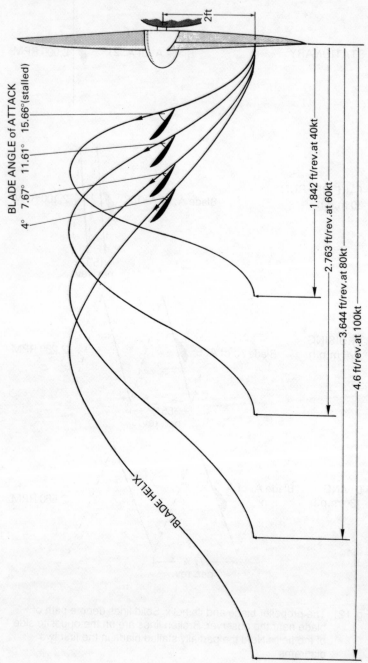

Fig. 122 Effect of speed on blade angle of attack at a constant 2200 RPM (not drawn to scale in the interest of clarity).

BLADE ANGLE of ATTACK

4°   7.67°   11.61°   15.66°(stalled)

2ft

1.842 ft/rev.at 40kt

2.763 ft/rev.at 60kt

3.644 ft/rev.at 80kt

4.6 ft/rev.at 100kt

BLADE HELIX

and which is adjusted on the throttle. The **Manifold Pressure Gauge** is used for correct setting of the throttle.

Section 5 of the aircraft manual will include a table of power settings under the title 'Cruise Performance'. Figures are quoted for different altitudes and temperature conditions. There will be several different ways of obtaining, say, 75 per cent power. The pilot could use 2500 RPM and 24 in. manifold pressure or 2400 RPM with 25 in. manifold pressure. At lower RPM it may not be possible to obtain 75 per cent power.

To avoid over-stressing the engine in a manner that may be likened to driving a car at low speed in top gear up hill, the following engine handling procedure should be adopted.

INCREASING POWER: Increase the RPM first, then open the throttle.
DECREASING POWER: Throttle back first, then reduce RPM.

The sequence can be remembered by the couplet:

Rev UP and Throttle BACK

The Constant Speed Unit

Constant speed propellers vary in design but generally pitch is changed by hydraulic pressure, using engine oil and a separate pump. Some pitch changing units use large coil springs and/or centrifugal weights to move the pitch in one direction, relying on oil pressure to make the opposite pitch change. The design details need not concern the PPL holder.

Pressure oil to and from the pitch changing mechanism, which is housed within the propeller hub, is controlled by a valve in the constant speed unit. The valve is itself moved to finer pitch, coarser pitch and pitch lock positions by a governor unit which is rotated by the engine. **Bob Weights**, which have a natural tendency to fly outwards as the governor is rotated, are held in the neutral position by a spring. The tension of this spring may be adjusted by the pilot by using the propeller lever.

If, for any reason, the engine attempts to increase speed above the RPM set by the pilot the bob weights will fly outwards and lift the oil valve, allowing hydraulic pressure to enter the propeller hub and move the blades to a coarser pitch when the RPM will return to the original setting, the governor spring will move the bob weights back to neutral and the valve will close, hydraulically locking the blades into the new pitch.

Should the engine lose RPM in, for example, a climb or slight nose-up attitude (leading to a reduction in speed) the bob weights would no longer be able to balance the spring pressure; they would move inwards, the valve would lower and allow the blades to adopt a finer pitch. RPM would be restored, the bob weights would return to the neutral position and the valve would hydraulically lock the blades at the new, finer pitch.

When the throttle is opened the CSU (constant speed unit) prevents any change in RPM in the manner just explained but the blades absorb the additional power by taking on a coarser pitch.

Changes in RPM are only possible when the governor spring tension is adjusted via the pilot's propeller lever. A simplified constant speed propeller unit is illustrated in Fig. 123.

Malfunctions

A fault in the constant speed unit could cause one of the following situations.

1. *Propeller fixed in fully fine position.* In this case the amount of throttle will have to be limited, otherwise the engine will over-speed and RPM may exceed the red line. Such a condition is sometimes called an 'overspeed prop'. It may be difficult to fly at other than low speed without exceeding the maximum RPM and an early landing is essential.
2. *Propeller fixed in fully coarse position.* In this situation the aircraft will cruise normally but if, during the landing approach, it is necessary to execute missed approach action some aircraft will be unable to develop enough power for the climb, particularly when fully loaded.

Usually propeller faults of this kind, which are very rare, result from failure of the propeller governor. It is sometimes possible to rectify the situation by closing the throttle, moving the propeller lever to the fully coarse position, reducing the airspeed and then adding power. If there is any propeller control available the propeller lever should be slowly moved to re-set RPM.

Number of Blades

For engines of up to 250 HP it is usual to fit two-blade propellers. Although additional power can be absorbed and converted into thrust by increasing the blade area usually this entails accepting a

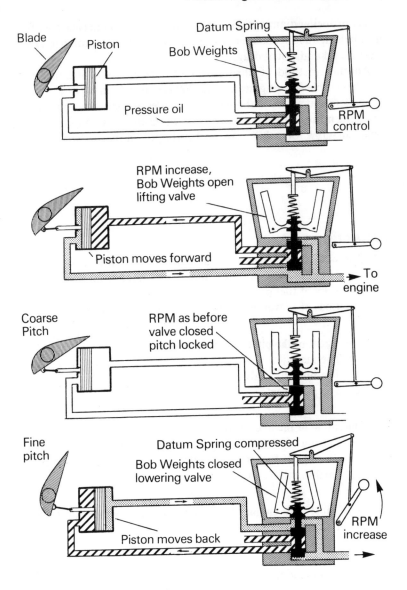

Fig. 123 Constant-speed propeller (diagrammatic). The first three drawings refer to the sequence described in the text. In the bottom illustration higher RPM have been selected. Hydraulic leads to the propeller are in fact drillings in the propeller shaft.

243

6   Piston Engines and Propellers

larger propeller diameter. To provide sufficient ground clearance for the blade tips a longer undercarriage will be necessary leading to extra weight and additional drag.

Another method of absorbing higher power is to increase the number of blades. Engines of 285–700 HP usually drive three-blade propellers and above that power four blades are used. As a means of reducing noise levels there is a move towards the slow-running multi-blade airscrew for public transport turboprop aircraft.

Questions

1 Too weak a fuel/air mixture in a piston engine will cause:
(a) High fuel consumption and black smoke from the exhaust,
(b) Loss of fuel pressure,
(c) Loss of power, overheating and possibly detonation.

2 Movement of the pilot's throttle control alters:
(a) The main jet,
(b) The butterfly valve,
(c) The power jet.

3 Operation of the idle cut-off stops the engine by:
(a) Cutting off fuel supply to the slow running jet,
(b) Cutting off fuel supply to the carburettor float chamber,
(c) Earthing the ignition.

4 With most aero engines in popular use each cylinder provides a power stroke:
(a) Every two engine revolutions,
(b) Every four engine revolutions,
(c) Every engine revolution.

5 In a piston engine, in-going mixture and out-going burned gases are controlled by:
(a) The carburettor heat control,
(b) The mixture control,
(c) The inlet and exhaust valves.

6 Two separate ignition systems are fitted to an aero engine for the purpose of:
   (a) Safety in the event of an ignition failure,
   (b) To provide better combustion and safety in the event of an ignition failure,
   (c) To assist engine starting.

7 When the ignition is switched 'off' the magnetos are prevented from generating sparks because:
   (a) The magnetos are earthed to the engine,
   (b) The battery has been switched off,
   (c) The plug leads have been disconnected from the magnetos.

8 When the tanks are in the mainplane of a low-wing design fuel starvation in the event of a failed mechanical fuel pump is safeguarded by:
   (a) Gravity feed,
   (b) The throttle-operated accelerator pump,
   (c) An electric fuel pump.

9 Carburettor icing is of three types:
   (a) Fuel evaporation, impact and throttle ice caused by adiabatic cooling,
   (b) Rime, hoar frost and glazed ice,
   (c) Fuel evaporation, rime and glazed ice.

10 The development of carburettor icing may be recognized by:
   (a) A gradual decrease in RPM, rough running and eventually complete loss of power,
   (b) Severe misfiring and fluctuation of the RPM indicator,
   (c) Sudden loss of power.

11 Before the first flight of the day it is good practice to check the ignition is off, then turn over the engine by hand before starting. Why?
   (a) To fill the cylinders with mixture and make the engine ready for a cold start,
   (b) To break the oil film adhesion and so reduce the load on the starter, to check the cylinder compressions and to check for hydraulicing,
   (c) To prime the oil system.

12 Assuming little or no wind and no met or air traffic restrictions a pilot wishing to fly for maximum range must adopt the following procedure:
   (a) Select a low altitude and fly at the minimum power setting for level flight,
   (b) Fly at a high altitude in weak mixture and at the minimum power setting for level flight,
   (c) Climb to an altitude where full throttle is required to maintain an indicated best L/D speed and weaken the mixture.

13 During take-off nosewheel aircraft have less tendency to swing than tailwheel types because:
   (a) There is no gyroscopic or asymmetric blade effect from the propeller and nosewheel undercarriages have good directional stability,
   (b) The pilot has a better view ahead,
   (c) Slipstream effect cancels torque effect.

14 When the power setting is altered there is a tendency to yaw due to:
   (a) Offset fin or fixed rudder trim,
   (b) Gyroscopic effect,
   (c) Slipstream and torque effect.

15 When an aircraft is fitted with a fixed-pitch propeller a change in airspeed will alter the engine RPM. Why is this?
   (a) Because of asymmetric blade effect,
   (b) Because an increase in airspeed will remove some of the load from the propeller and allow an increase in RPM while a decrease in airspeed has the reverse effect,
   (c) Because changes in airspeed alter the amount of air passing through the induction system and this affects engine power.

16 If a coarse-pitch propeller (fixed) ensures a high cruising speed why are they not always fitted to light aircraft?
   (a) Because of noise limitations,
   (b) To avoid high fuel consumption,
   (c) Because a poor take-off performance results from this type of propeller.

17 **Why do some propellers have three blades instead of two?**
   (a) To absorb high engine power without having to resort to propellers of large diameter,
   (b) To provide safety in the event of blade failure,
   (c) To improve take-off performance.

18 **Some engines are fitted with fuel injection systems in place of a carburettor:**
   (a) To prevent condensation and vapour locks in the fuel system,
   (b) To increase the power of the engine,
   (c) To improve fuel economy and reduce risk of icing.

19 **A fine-pitch propeller performs better than one of course pitch;**
   (a) During the cruise,
   (b) During the take-off,
   (c) During the take-off and climb.

20 **The tendency for an aircraft to roll in the opposite direction to propeller rotation is known as:**
   (a) Asymmetric blade effect,
   (b) Torque reaction,
   (c) Lateral trim.

21 **At constant engine RPM blade angle of a fixed-pitch propeller:**
   (a) Reduces as the airspeed increases,
   (b) Increases as the airspeed increases,
   (c) Remains the same irrespective of airspeed.

22 **It is bad practice to run up an engine while parked on an area of loose stones:**
   (a) Because there is a risk that the brakes may not hold,
   (b) Because stones may be lifted to damage the propeller,
   (c) Because a stone could enter the engine induction system and be ingested into a cylinder where it would cause damage.

Chapter 7
Aircraft Systems

So far the chapters dealing with aircraft in this manual have been confined to the airframe, engine and propeller. These major components would, on their own, be no more practical than a house without electricity, running water and some form of heating.

The various services fitted in aircraft are known as **Systems** and even a simple trainer will have the following:

Fuel
Electric
Pitot/static supply
Vacuum
Cabin air heater.

Larger and more complex aircraft will also employ:

Hydraulics
Cabin pressurization.

Like much of the technical information in this book the following text is limited to the needs of pilots; it is not intended for engineers under training. However, although many motorists know little about their cars private pilots, even those who aspire to no more than a little pleasure flying within the local area, must have a working knowledge of the systems that make it possible to fly the aircraft.

Fuel System

Fuel tanks are usually made of light alloy and they may be located one in each wing or, in the case of single tank systems, in the fuselage below the baggage shelf which is behind the cabin seats.

A **Fuel Selector** (similar in design to a domestic water tap) is provided. In some aircraft it can be moved from OFF to LEFT TANK or RIGHT TANK although the most widely used trainers have a system which allows both wing tanks to drain into a common point. In such cases the fuel selector will be of the two-position type, i.e. either ON or OFF.

When individual tanks may be selected some aircraft have a fuel selector which points to the fuel contents gauge of the tank in use. Another safety device is a button which must first be depressed before the fuel can be turned OFF, thus preventing accidental selection which would starve the engine of fuel.

Although most aircraft employ a combined, key-operated ignition/starter switch some have a separate starter button which is hidden by the fuel selector until it is turned ON, an arrangement which guards against the possibility of attempting a take-off with the fuel turned OFF.

Fuel Capacity

Because of design restrictions it is usually not possible to use all the fuel in a tank and a small quantity always remains after it has ceased supplying the engine. Consequently two fuel capacity figures are usually quoted, **Total** and **Usable**. The amount of unusable fuel is in the region of one gallon per tank in the case of light trainers and this is included in the empty weight of the aircraft.

Each tank is equipped with a car-type fuel contents transmitter, employing a float on a moving arm that is attached to a variable resistance device. The contents transmitters are connected to fuel gauges on the pilot's instrument panel.

Fuel Tanks

The filler caps may be fitted with locks but whatever the design it is essential to ensure that after filling the cap is properly replaced and twisted fully home into the locked position.

To prevent an airlock developing as fuel is drawn from the tank a **Vent** pipe is fitted and it usually incorporates a simple device to ensure that air, not fuel, may pass through the pipe. In some aircraft tank venting is provided in the filler cap.

Fuel Straining

When an aircraft is allowed to stand over a period of time with partly filled tanks there is a tendency for moisture in the air trapped above fuel level to condense on the tank walls. If water does enter the carburettor jets, being more viscous than fuel, it will fail to pass through the very fine orifice provided for atomizing and the engine will be starved of mixture.

To rid the tank of water or other impurities a small sump is built into the under surface. Water, being heavier than fuel, sinks to the bottom of the tank and a **Fuel Strainer** is provided in the sump through which fuel or, when present, condensed water, etc., may be drawn.

There may be an additional fuel strainer at the lowest part of the fuel system. It will usually be located within the engine bay.

Precautions to be Adopted while Straining Fuel

The common practice of pushing up the fuel strainer valve and allowing it to pour onto the ground is pointless and potentially dangerous because it does not tell the pilot if contamination is present.

Some aircraft have a small jar provided for the purpose. A thin rod extends, upwards through the centre and this is pushed into the strainer, opening a spring valve and allowing contents from the bottom of the tank to pour. When water is present it will be seen in the jar, eventually with coloured fuel floating on top. Some fuel strainers are opened by pushing up a ring and allowing them, to drain into a small glass jar.

Whatever the type of strainer care must be taken to ensure that it has properly closed after checking for fuel contamination.

Fuel straining must not be attempted when the fuel selector is OFF. If, for any reason, the tank vent is blocked an incorrectly closed strainer would probably stop pouring and mislead the pilot into believing that it is properly seated. When the fuel is turned ON and pressures within the fuel system equalized petrol could then issue from the strainer and lead to the aircraft running out of fuel in flight.

Refuelling Precautions

When the airlines of the world flew large piston-engine aircraft pilots of light aeroplanes had to ensure that high octane fuel was not added to the tanks in error. It is still important to ensure that the correct grade of fuel is used and this will usually be AVGAS 100LL (blue colour) or 100 (formally known as 100/130 and green in colour).

Aviation petrol is known as **AVGAS** but some flying training establishments have for some years operated successfully with motor fuel or **MOGAS**, the only recorded problem being a tendency for, some engine installations to cause vapour locks due to the fuel pump running hot.

Now that many aircraft are turbine powered (**Turboprop** or **Fanjet**) particular care must be taken to ensure that turbine fuel (**AVTUR**) is not put into piston-engine aircraft. Confusion has arisen in several cases because turbocharged aircraft had the word 'Turbo' painted on the side of the engine cowling and this led the refuelling crew to believe that the aircraft was turbine powered.

System Layout

Fig. 124 shows a typical fuel system of the kind likely to be fitted in a light trainer. It will be seen that from the tanks feed pipes lead to the fuel selector. From there, fuel is led to a fuel strainer which removes solid contamination that may have found its way into the tanks during refuelling.

At the strainer a pick-up point is provided for the primer (when one is fitted) and a main feed is then directed to the carburettor via suitable pumping arrangements.

Provision of Fuel Feed

Fitted to the engine is a diaphragm pump for the purpose of drawing fuel from the tank(s). When the aircraft is a high-wing design and the tanks are mounted in the wings fuel feed is, in addition, ensured by gravity.

Many aircraft are of low-wing type and in these the tanks are below the level of the carburettor, consequently provision must be made to ensure fuel supply in the event of mechanical fuel pump failure. In any case, without fuel pressure it would not be possible to prime the engine for starting. The problem is solved in a car by having a large capacity battery which is capable of turning over the engine until the mechanical pump has lifted fuel from the tank to the carburettor. Alternatively, there will be an electric fuel pump which supplies fuel pressure immediately the ignition is switched on.

Low-wing aircraft of modern design are fitted with an electric fuel pump which must be switched on before starting the engine. A **Fuel Pressure Gauge** is grouped with the other engine read-outs and this will confirm fuel pressure at all times, prior to starting, in flight when the electric pump is off and the engine is depending on its mechanical diaphragm pump for supply and so forth.

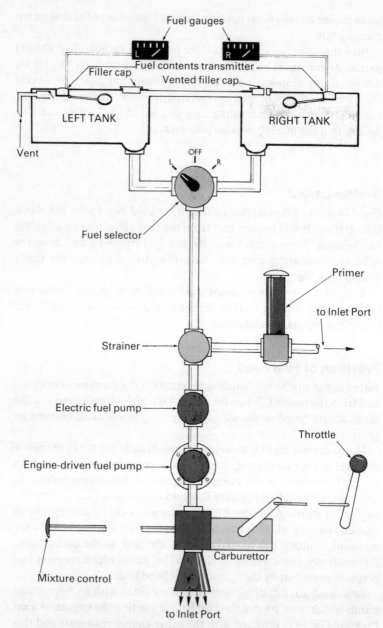

Fig. 124 Typical two-tank fuel system as fitted in a light aircraft. High wing designs, where the tanks are located at cabin roof level, usually dispense with an electric fuel pump since gravity feed is available to back up the engine-driven pump.

Use of the Electric Pump

Some pilots have a habit of switching off the electric fuel pump during the early stages of the climb, sometimes when the aircraft is below 500 ft. Such action totally defeats the purpose of fitting an emergency back-up pump. If the engine does fail shortly after take-off the most common cause is likely to be fuel starvation, and if the electric pump is switched off an appreciable number of seconds may be required for it to re-establish fuel pressure again. Without power seconds mean height loss in feet at a time when such losses can be dangerous. Often the pilot is tempted to turn back. Insufficient height leads to desperate attempts at tightening the turn while holding up the nose to conserve height. There can be only one outcome – the aircraft will spin in (see Exercise 12, *Flight Briefing for Pilots, Vol. 1*).

It is good practice to leave on the electric pump until the top of the climb.

Priming Pump

To provide rich mixture for cold starting cars are fitted with a choke which may be manually controlled by the driver or of the electric, automatically controlled type.

Aero engines, other than those adapted from car engines, use a priming pump instead of a choke. It takes the form of a small plunger-type pump which is located on the instrument panel. After fuel pressure has been obtained the plunger is released by twisting the pump knob through part of a turn. It may then be drawn back to suck in fuel then pushed forward, injecting fuel into the engine inlet ports. Several strokes of the pump may be necessary before it clears air from the system and takes in fuel. This will be felt as an increase in effort needed to work the pump.

The number of priming strokes needed for a cold engine start will depend on the temperature. The recommended number of primes will be given in the aircraft manual.

After priming and engine starting it is essential to ensure that the pump is correctly locked into its stowed position, otherwise fuel mixture could be seriously disturbed leading to rough running and excessive fuel consumption.

Electric System

Electric circuits fitted in light trainers may be 12 volt or 24 volt systems. They are required to supply:

Engine starting
Navigation and landing lights
Cabin lights
Instrument and radio lights
Anti-collision/strobe lights
Pitot heat (to prevent loss of the ASI and altimeter through ice)
Radio equipment
Some instruments and engine read-outs.

In some aircraft the electric system also works the flaps and more advanced touring aircraft sometimes have an electrically retracting undercarriage.

Aircraft Batteries

To operate the electric starter and other services there will be a 12 V or 24 V battery (according to system voltage). This is usually located ahead of the engine firewall and retained in a vented box. Weight being an important limitation, particularly in low-powered trainers, aircraft batteries are usually of low capacity, typically 14–25 amp/hour compared with 40–66 AH in cars. It therefore follows that when the engine is not running all non-essential equipment should be switched off until there is generating power. The practice of switching on the anti-collision beacon prior to starting is something that many light aircraft pilots have copied from the airlines. Large aircraft have adequate electric power, either from an external source or from within. Light trainers do not and, particularly on a cold day, one can pay for the luxury of copying the airline pilots by failing to start.

Of the electrical services listed at the start of this section engine starting makes the biggest demands on the battery and unless the current drain is replenished there would soon be no electric power.

Alternators and Generators

While older types of aircraft are fitted with an engine-driven generator most modern designs use **Alternators**. These are capable of providing electric current at lower engine RPM than generators. Generators provide **Direct Current** (DC) whereas alternators supply **Alternating Current** (AC) which has to be converted to DC (the operating system in light aircraft) by a simple diode-type rectifier.

The purpose of the alternator is:

1. To maintain the battery in a fully charged state.
2. To support the electrical services in flight.

Steady voltage is controlled by a **Voltage Regulator** and to protect the system there is an **Over Voltage Relay**. Should this operate a red warning light, sometimes labelled HIGH VOLTAGE, will illuminate. Steps to be taken will be described in the next section.

Most light aircraft are fitted with standard car industry alternators and starters. Usually the alternator is belt-driven via a pulley mounted behind the propeller spinner.

Battery Master Switch

To ensure that the battery is not inadvertently left supplying services on the ground while the aircraft is parked for any length of time a **Master Switch** is fitted. In most aircraft of modern design this is of the **Split-rocker** type, i.e. two thin rocker switches mounted side-by-side. They are coloured red and the left half is labelled ALT (Alternator) with the right half marked BAT (Battery). Like all aircraft switches the battery master is ON when it is UP.

In normal use both halves of the switch are operated together and when the master is OFF all electric services are disconnected from the battery with the exception of the clock.

The purpose of splitting the master switch is to allow operation of the electric services via the battery in flight while the alternator is being re-set following a HIGH VOLTAGE warning. Should the red light appear this indicates that the alternator has, for some reason, automatically shut down. Attempts to re-set the alternator should be made by turning off the ALT section of the master switch for a few seconds and then turning it on again. If the red light continues to show there has been an alternator malfunction. This will be confirmed on the **Ammeter** which will indicate a lack of charge.

All non-essential electric load must be switched off to save the battery (some aircraft need electric power for lowering the flaps) and an early landing should be made.

Ammeter

This instrument shows the flow of current in amperes from the alternator to the battery. It will register the highest current flow when the battery is at low charge and there will be a no charge indication when the battery is up to its full capacity.

When the aircraft is on the ground with the engine stopped and various electric services are switched on the ammeter will register a discharge which becomes heavier as more and more load is applied by turning on radio, lights, pitot heat, etc. The alternator is of sufficient output to supply all the radio equipment and other services in flight but if, at any time the ammeter shows a discharge although the HIGH VOLTAGE light is not illuminated this indicates that, for some reason, the alternator cannot meet the current demand and the pilot should switch off non-essential items. Such an occurrence must be reported on landing.

Supplying the Electric Services

Fig. 125 shows a simple electric circuit of a type likely to be fitted in a light trainer.

It will be seen that electric current, either from the battery or the alternator is fed to a main supply contact known as a **Bus Bar**. This is the 'Positive' side of the circuit while 'Negative' is conducted throughout the aircraft via the airframe.

The various services – radio, instruments, lighting, etc. – are connected to the bus bar and are protected from overload by:

1. Thermal Overload **Circuit Breakers** which pop out under overload conditions, or
2. Fuses similar to those used in a car.

Part of the pre-starting check is to ensure that all circuit breakers are 'in'. If, at any time, a circuit breaker has popped out there may have been a temporary overload and supply to that particular circuit may be resumed simply by pushing it in again. When the fault persists then the equipment being supplied will most likely have suffered a major fault and it should be switched off.

External Power Plug

Mention has previously been made of the trolley-acc which may be used at some airfields for the purpose of saving the aircraft's battery during servicing and while starting. An external plug is required for the purpose and these are usually fitted by the manufacturers as an option.

When an external power socket is fitted there will be a battery contactor solenoid which automatically switches the aircraft circuit to the trolley-acc while it is connected and re-connects the internal battery when the plug is withdrawn. There is a separate fuse for this optional facility and this is mounted on the engine firewall.

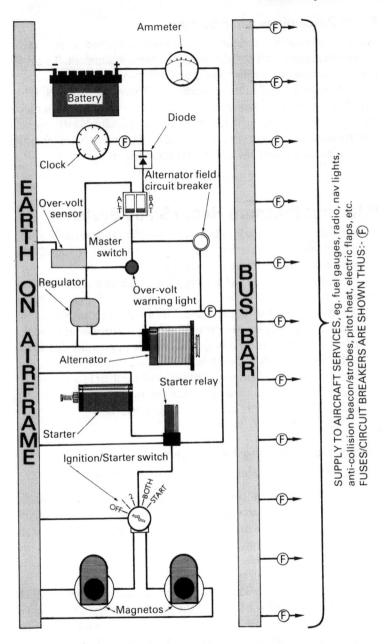

Fig. 125 An electric circuit of the type fitted to light aircraft.

Radio Precautions while Starting

When the engine fires and the alternator comes 'on line' there is a risk of current surge which could damage the radio equipment. To avoid possible damage all radios should be OFF prior to starting.

While each item of radio equipment will have its own ON/OFF switch, in some cases incorporated with the volume control, a more convenient arrangement is the provision of a **Radio Master Switch** which avoids the need to switch on or off perhaps six or more separate radio items every time the aircraft is entered or left.

Pitot and Pressure Head/Static System

To provide the airspeed indicator, altimeter and vertical speed indicator with suitable air supply careful steps are taken to obtain the sample away from airframe disturbances.

The altimeter and the VSI are supplied with **Static** air, i.e. air at a pressure not distorted by aircraft movement which represents the prevailing pressure for that altitude. The airspeed indicator, on the other hand, requires two samples of air for comparison, static and the same static air that has been subjected to a rise in pressure due to the aircraft's movement (i.e. **Dynamic** air).

Because the ASI, altimeter and VSI depend for their function on air pressures they are known as **Pressure Instruments**. These and the other aircraft instruments are described in Chapter 9.

Methods of Obtaining Air Samples

Two main pressure sensing methods are used in light aircraft:

1. **The Pitot Head**, a combined pressure/static device suitably located on the aircraft in a position away from airflow disturbance. It consists of an open-ended tube which samples pressure air. Surrounding it is a concentric tube which is blocked at the front. A series of small holes are drilled along the sides to admit air at static pressure.

   The pitot head often incorporates an electric element which may be switched on to avoid the build-up of ice which would result in the loss of all pressure instruments in some cases and distorted reading in others according to the extent of icing.

2. **Pressure Head/Static Vent.** An alternative system makes use of an open ended pressure head which, like the pitot tube, may

incorporate a heater. Static air is obtained from a static vent which takes the form of a small plate mounted on the side, or both sides, of the fuselage. The plate has a small hole to admit static air.

Pressure Instrument System

The pressure/static sources previously described are connected by a system of light alloy tubes to the instruments.

Over a period of time moisture can collect in the system and a drain valve is often provided. In case there is a sudden blockage of the static source it is usual to have an **Alternative Static Source** valve which can be operated from the cabin. This will restore the altimeter and VSI following loss of readings due to a blocked static source. It may also reinstate the ASI unless its pressure source is blocked. Because the air sample is from within the cabin there are likely to be small inaccuracies in the readings of these instruments.

To protect the pitot/pressure head from the intrusion of water or insects a cover is provided and this should be tied in place while the aircraft is parked for any length of time. The **Pitot Cover** is usually coloured bright red to remind pilots that it must be removed before take-off.

A typical pressure/static system is illustrated in Fig. 126.

Vacuum System

The attitude indicator and the direction indicator each incorporate a vacuum-driven gyro. These instruments are described in Chapter 9.

Most light aircraft of current design obtain vacuum supply from an engine-driven **Vacuum Pump** which draws air from the instrument cases so that filtered incoming air can be directed through suitable jets onto the gyros. The normal vacuum suction is 4.6 to 5.4 in. of mercury and a vacuum relief valve is incorporated in the circuit to maintain these values. A small suction gauge will be located on the instrument panel and this must be checked during the engine run-up.

The vacuum system is illustrated in Fig. 127.

Some light trainers equipped to minimum standards utilize a **Venturi Tube** in place of a vacuum pump. This trumpet-shaped tube is positioned outside the aircraft where it is exposed to the airflow. As air is forced through the restriction of the venturi a decrease in

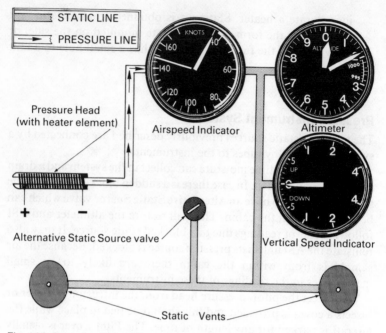

Fig. 126 Pressure Head/Static system. The pressure head is usually located under one wing and the static vents are positioned on each side of the fuselage (some aircraft have only one).

pressure occurs at the throat where a tube connection supplies vacuum to the instruments.

The disadvantages of venturi tubes are increased drag, risk of ice in certain conditions and the fact that until the aircraft is flying there is insufficient vacuum to spin the gyros and provide reliable attitude/direction readings. A better method is the **Exhaust-driven Venturi** which is sometimes fitted to very light trainers. In this system the engine exhaust passes through a venturi tube which in turn generates suction for the gyro instruments.

Cabin Air and Heater System

Although the outside air temperature drops with altitude, being some 15° cooler than sea level at 5000 ft, aircraft with large areas of clear plastic can become very hot in the cabin when the sun is shining.

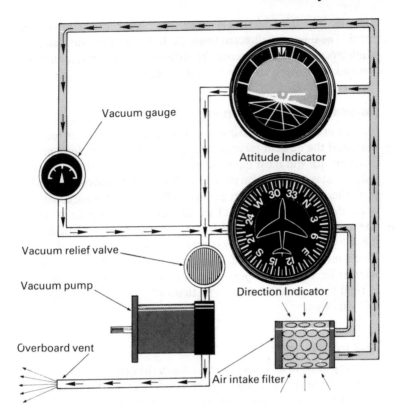

Fig. 127 Vacuum System. The Turn and Balance or Turn Co-ordinator will be electrically driven to cater for possible failure of the vacuum instruments.

To some extent the temperature may be reduced by fitting tinted panels inside the canopy but good ventilation is essential if pilot and passengers are to remain comfortable in flight. Cold air is introduced to the aircraft through small scoops which may be situated in the wing roots or alternatively, the cabin sides. Short lengths of ducting guide the air to the cabin via adjustable outlets which may be closed or progressively opened in much the same manner as car vents.

Another air inlet source is directed through a heat exchange muffler which surrounds part of the exhaust system and a mixing trap valve, operated by a temperature control lever in the cabin, determines the proportion of cold to hot air that will be supplied. Another valve apportions the warm/hot air to the cabin or defrosting ducts let into the **Glareshield** which sits on top of the instrument panel. The ducts are directed towards the lower inner surface of the windscreen for the purpose of demisting or, when conditions demand, defrosting areas of the windscreen.

De-frosting the Windscreen

If an aircraft has been flying in below-zero temperatures for some time, and the descent takes it through an area of thin cloud or even moist air, ice will most likely form on the outside of the windscreen (hoar frost is described on page 114, Chapter 3, Meteorology). In bad cases there can be sufficient ice to totally obscure forward vision.

To clear the windscreen of ice the selector should be moved to direct all air to the screens with the temperature control adjusted to the fully HOT position. Two small areas will be cleared on the windscreen and these will enlarge as the aircraft descends into warmer air. During this period the aircraft should be turned from left to right through a few degrees so that fuller use can be made of the side windows.

Risk from Fumes

If, at any time, there is a smell of fumes in the cabin it is possible that the exhaust may be leaking through a faulty heat exchanger although this is very rare. However, the risks of carbon monoxide poisoning are well known and if there is any doubt about the origin of the fumes the cabin heater should be turned off immediately and the fresh air vents should be fully opened.

More Complex Aircraft (*This section is not part of the PPL Syllabus*)

The systems so far described will be found in the majority of modern light training aircraft. At a later stage the private pilot may fly advanced single or twin-engine designs which have additional systems. These are now briefly described.

Hydraulic System

Although both flap operation and undercarriage retraction and lowering is often made with electric actuators, in some aircraft hydraulics are used.

The service to be operated is moved by a **Hydraulic Jack**, a device that resembles a bicycle pump in many respects. Oil under pressure is directed to either side of the piston that moves within the jack. A selector valve directs pressure oil to the UP or DOWN side of the jack via a system of hydraulic pipes or **Lines** as they are often called. As the piston moves under the influence of pressure oil that on the other side of the piston is forced out down a **Return** line to the **Hydraulic Reservoir**.

Hydraulic pressure may be generated by an engine-driven pump or there may be a self-contained **Electro-hydraulic Power Pack** which consists of an electric motor driving a hydraulic pump. Sometimes the motor is reversed to provide UP and DOWN movement of the undercarriage and/or flaps.

Pressurization

At an early stage of training the student will become aware of the advantages to be gained by flying at higher altitudes. Even a low-powered trainer will offer a faster and more economical cruising speed for any given power setting when flown at, say, 8000 ft rather than 2000 ft. This is because while fuel flow in gallons per hour remains unchanged with height for any power setting, TAS becomes increasingly greater than IAS as height is gained. The net result is an improvement in air miles per gallon.

The other important advantage to flying high is that most or even all of the weather may be avoided leading to a smoother flight although cruising altitudes of at least 30,000 ft are required to clear the tops of cumulonimbus clouds which can reach even greater heights in some parts of the world.

In so far as the aircraft is concerned the limiting factor while gaining height is the availability of power. To a considerable extent the introduction of turbocharged engines (p. 230) has enhanced the climbing and altitude performance of piston-engined aircraft (although even the best of these power units cannot match the high altitude performance of a jet).

Aircraft capabilities are one thing; human limitations are another, and high altitudes bring with them certain physiological problems. First there is the discomfort that can result from larger than usual pressure changes. This can cause pain in the ears and even damage when the person affected has a cold. The other limitation is lack of oxygen.

7 Aircraft Systems

To cater for lack of oxygen, which becomes significant when flying at altitudes above 12,000 ft, some turbocharged aircraft have an oxygen system which entails the occupants wearing masks. Whether or not one is prepared to wear an oxygen mask there remains the problem of low pressure and the possibility of ear discomfort. To deal with these matters the **Pressurized** cabin was developed.

Cabin Altitude

To the student pilot, learning to fly on a simple, low powered trainer, talk of high speed cruising at 20,000 ft or more may seem to be of little relevance in a book of this kind. Nevertheless, there are several high performance light singles which have pressurized cabins and this introduction to the subject is certainly not out of place.

In a non-pressurized aircraft atmospheric pressure within the cabin is the same as that outside. By strengthening the cabin area and sealing all gaps in the structure **Bleed Air** from the turbocharger can be made to increase pressure inside, at the same time pumping in additional air. The maximum pressure difference between conditions inside the aircraft and the atmosphere outside is known as the **Pressure Differential**. In a light single-engine aircraft it will be in the region of 3 to 4 lb/sq in. Jet aircraft, which are capable of flying at 40,000 ft or higher, have a pressure differential in excess of 8 lb/sq in.

The advantages of pressurization are perhaps best illustrated by comparing the difference between real altitude and the apparent altitude within the cabin (known as **Cabin Altitude**). The figures relate to a well-known pressurized light single.

Aircraft altitude (ft)	Cabin altitude (ft)
7000	sea level
10,000	800
14,000	5500
18,000	8500
23,000	12,100

It will be noted that up to a cruising altitude of 7000 ft the cabin remains at sea level and the aircraft may be climbed to an altitude of 23,000 ft before the cabin reaches the practical limit for human comfort.

Provision is made to heat or cool the air being supplied to the cabin by the turbocharger and to prevent over-pressuring there is a **Dump**

264

Valve, which is similar in concept to the safety valve that was fitted to steam engines.

Operating the Pressurization

While systems vary from one aircraft type to another the following brief description will illustrate the principles involved.

Pressurization is managed on a small panel which contains an ON/OFF switch. Next to it is a **Cabin Altitude Selector** and this is used to select the altitude at which pressurization will begin. Scales are provided for cabin altitude and an adjacent scale gives the equivalent aircraft altitude for the system, the difference between the two figures being dependent upon the capabilities of the pressurization in terms of pressure differential.

The system is supported by two instruments:

1.   **Cabin Rate of Climb**
2.   Combined **Cabin Altitude** and **Cabin Differential Pressure**.

Before starting the engine the pressurization controls should be set. The dump valve is closed, the pressurization switch is turned on and the cabin altitude selector should be set to 1000 ft above the departure or arrival airfield, whichever is higher.

Some manufacturers recommend that during take-off the throttle should be advanced slowly to avoid a sudden increase in cabin pressure which may cause some 'ear popping'. Alternatively, the system may remain switched off until after the climb has been established. Pressurization starts when the aircraft has climbed through the level set on the cabin altitude selector.

A **Cabin Altitude Warning Light** is fitted and this will illuminate if at any time the aircraft climbs to a level where the system is unable to keep the cabin below the equivalent of 12,500 ft. When any changes are made on the cabin altitude selector while in flight these must be handled very slowly to avoid sudden pressure changes. These could affect occupants with pressure sensitive ears.

Since cabin pressure is supplied by the turbocharger it follows that the throttle must not be fully closed during the descent. Immediately prior to landing the cabin pressure differential should be checked to ensure that it is at zero; most aircraft are not cleared for landing while pressurized. As a precaution, the dump valve must be opened after landing and before opening the cabin door or windows. This will ensure that no pressure differential remains in the cabin.

Questions

1 The Aircraft Manual quotes fuel capacities. Amount of fuel available, assuming the tanks have been filled, is:
   (a) The sum of maximum fuel in all tanks,
   (b) Usable fuel,
   (c) Total fuel.

2 Air locks in a fuel system are prevented by:
   (a) Fitting fuel strainers,
   (b) Providing vent pipes,
   (c) Fitting an electric fuel pump.

3 Fuel strainers are provided:
   (a) To check for fuel contamination (water, impurities etc.),
   (b) To remove air locks in the fuel system,
   (c) To check that fuel is present in the tanks.

4 An electric fuel pump is fitted in some aircraft:
   (a) To supplement the mechanical pump and guard against its failure,
   (b) To assist in drawing fuel to the engine for starting,
   (c) To remove air locks.

5 A priming pump is fitted in some aircraft for the purpose of:
   (a) Supplying fuel to the induction system prior to a cold start,
   (b) Raising fuel from the tank of a low-wing aircraft to the mechanical fuel pump,
   (c) Providing a supply of fuel in the event of engine pump failure in flight.

6 The two halves of a split master switch:
   (a) Relate to each battery,
   (b) Relate one to the battery and the other to the alternator,
   (c) Relate one to the battery and the other to the radio.

7 Because light-aircraft batteries are of limited capacity, during starting it is good practice to:
   (a) Turn off all other load until the engine is running,
   (b) Always use an external battery source,
   (c) Make the first start of the day by hand-swinging the propeller.

8 Most modern light aircraft obtain electric current in flight:
  (a) From the battery,
  (b) From a generator supplying DC current,
  (c) From an alternator supplying AC through a diode rectifier.

9 To ensure a steady voltage over a wide range of engine speeds and electric loads the electric system is fitted with:
  (a) An Over-Voltage Relay,
  (b) A Voltage Regulator,
  (c) An Idle Cut-off.

10 After flying for a short while the ammeter ceases to indicate a flow of current from alternator to the aircraft's battery because:
  (a) The battery is flat,
  (b) The battery is now fully charged,
  (c) The alternator has failed.

11 Most of the electric circuits in a modern aircraft are protected by:
  (a) Car-type capsule fuses,
  (b) Individual switches,
  (c) Thermal overload circuit breakers which automatically pop out under abnormal load and isolate the circuit.

12 A static vent is fitted:
  (a) To provide fresh cabin air,
  (b) To avoid air locks in the fuel system,
  (c) To supply the altimeter, airspeed indicator and VSI with samples of static air.

13 A Pitot Head is:
  (a) A combined pressure and static tube located in the free airstream,
  (b) A tube sampling pressure air which is located in the free airstream,
  (c) A tube with a restriction which supplies vacuum to the gyro-operated instruments.

14 To guard against loss of altimeter and other readings in the event of a blockage in the pressure instrument system aircraft are fitted with:
  (a) Pitot heaters,
  (b) An Alternate Static Source Valve,
  (c) Static Vents.

15 Although it is usual to fit an electrically-driven turn needle or turn co-ordinator, the gyros in the attitude and direction indicators are rotated by:

(a) Pressure air from the pressure head,

(b) Pressure air from the Pitot head,

(c) Vacuum from the engine-driven vacuum pump.

16 Heat for warming the cabin and de-frosting the windscreen is provided by:

(a) The carburettor heat system,

(b) A heat exchanger built around the engine exhaust system,

(c) The Pitot heater.

Chapter 8
Aircraft Performance, Weight and Balance

So that a flight can be planned the pilot must be able to obtain information about:

1. Take-off Performance. When the far hedge is beginning to fill the vision is no time to find out that, because it is fully loaded, the aircraft needs a longer take-off run than is available at that airfield.

2. Fuel, Time and Distance to Climb. This informs the pilot how much time will be required to reach the planned cruising level, how far the aircraft will have travelled along the route by the time it is at the required altitude and how much fuel has been used.

3. Cruising Performance. Essential for flight planning because only by knowing how much fuel is required can the decision be taken on whether or not an intermediate landing will be necessary. Also, by flying a little slower time could be saved by avoiding a landing for refuelling part way along the route.

4. Landing Performance. Although this is an academic consideration in so far as light trainers are concerned, the larger touring aircraft can, when landing at near-maximum weight, require more runway length than is sometimes available at the smallest airfields.

Aircraft Performance

The information now described will be detailed in Section 5 of the aircraft manual and the following text is intended to introduce student pilots to methods of presentation.

Take-off Performance

While the small, two seat trainer can under normal circumstances be operated out of most airfields it is sometimes erroneously believed that more advanced single and twin-engine aircraft can be flown out of small airstrips regardless of conditions. On too many occasions accidents, some of them serious, have resulted because it was not realized that:

(*a*) There are a number of factors materially affecting the take-off performance of an aircraft, and

(*b*) published **Take-off Run** figures are very much shorter than those for **Take-off Distance** which is from the start of the take-off roll to a position 50 ft above ground level.

Factors affecting take-off performance

1. Wind conditions; strength, direction and turbulence
2. Percentage of Max. Total Weight Authorized at the time
3. Temperature
4. Airfield elevation
5. Type of surface
6. Surface gradient
7. Use of flap
8. Obstructions during climb out
9. General condition of the aircraft.

1. Wind

A tailwind that is 10 per cent of the lift-off speed will add some 20 per cent to the nil wind take-off run while a similar headwind will shorten the run by 10 per cent. The moral is clear; always take off into wind whenever possible. Crosswinds will naturally reduce the headwind component.

2. Weight at Time of Take-off

Additional weight increases the take-off speed and at the same time decreases the rate of acceleration. In round figures 10 per cent increase in weight increases the lift-off speed by 5 per cent and adds 20 per cent to the length of take-off run.

3. Temperature

This directly affects air density which in turn will affect engine power and the amount of lift at any particular speed. In fact when the temperature is 10°C above ISA there will be a 10 per cent increase in take-off run.

4. Airfield Elevation

Air density decreases with altitude, therefore altitude is similar in effect to temperature. For example there will be a 10 per cent addition to the take-off run for each 1000 ft of airfield elevation. The relationship between temperature and altitude is such that before proper use can be made of the take-off performance information printed in aircraft manuals both factors must be known.

5. Airfield Surface
Most take-off run/distance figures quoted in the aircraft manuals assume operation from a hard, level runway. Normal dry grass will add approximately 15 per cent to the ground roll while long, wet grass or very soft ground can add 25 per cent to the run or even prevent lift-off speed being attained. Soft field take-off techniques are described on page 166 of Volume 1 in this series.

6. Surface Gradient
An uphill slope of 2° will extend the take-off run by about 10 per cent.

7. Use of Flap
Contrary to popular misconceptions not all aircraft respond to use of flap for take-off. Some types may even display a deterioration in take-off and initial climb performance when flaps are used. The subject is a complex one but the simple answer to the question of whether or not to use flap is to consult the Owner's/Flight/Operating Manual for the aircraft.

8. Obstructions During the Climb Out
Lifting off in safety is only part of the task. When high trees, power cables etc., must be cleared during the initial climb incorrect use of flap or adoption of an incorrect climbing speed can degrade both rate of climb and obstacle clearance. Take-off distance (i.e. to 50 ft a.g.l.) information in the manual is of prime importance under this heading.

9. General Condition of the Aircraft
With age engine performance declines and the cumulative effect of small airframe imperfections becomes measurable. It therefore follows that when consulting take-off graphs or performance figures in aircraft manuals it would be prudent to allow, say, 10 per cent safety margin to the figures quoted unless the aircraft is known to be in new, or almost new, condition.

The cumulative effect of all these factors can be very considerable and the still air take-off run for an aircraft normally requiring 800 ft or so could be extended to almost 3000 ft.

8 Aircraft Performance, Weight and Balance

Take-off Distance Information

Section 5 of the aircraft manual will give figures for both the Take-off Run and Take-off Distance. The latter is of more practical use because it assumes the aircraft to have arrived at a point 50 ft above airfield level (known as the **50 ft Screen**).

Some manuals provide take-off performance figures with and without the use of flap but in either case the lift-off, i.e. V_r or Rotate speed will be quoted. The information may be in the form of graphs or there could be a table of figures set out to give both take-off roll and take-off distance against these variables:

1. Pressure altitude (a longer roll/distance will be required when taking off from airfields situated at high altitudes).
2. Temperature above standard for any particular airfield level (an increase in temperature reduces air density and that decreases both engine power and lift, so causing the aircraft to take a longer run while taking off.

There will also be a statement in words advising that distances will be decreased by a percentage for every 9 kt headwind and increased by the same percentage (usually 10 per cent) for each 2 kt tailwind. There is also a percentage increase (usually 15 per cent) when the take-off is made off grass instead of a hard surface. In the absence of temperature information the pilot may calculate the density altitude, a method of correcting the existing pressure for the effects of temperature. The density altitude then found may be related to the take-off performance tables when no allowance is provided for temperature.

Calculating Density Altitude

The **International Standard Atmosphere** (ISA) at sea level is defined as having a pressure of 1013.2 millibars (29.92 in. mercury) and a temperature of +15°C. When the temperature at any altitude departs from ISA conditions (which assumes a temperature decrease with height of 6.5°C/1000 metres or approximately 2°C/1000 ft) the air density will likewise depart from the standard for that altitude. This means that on a hot day air density at the airfield (or any other altitude) will relate to a higher level and so affect both engine power and the amount of lift for any particular indicated airspeed. So although the pressure altitude, corrected by adjusting the pressure scale on the instrument, might read airfield elevation in so far as both engine and airframe are concerned, they perform according to the air density at the time. While this is catered for in the tables of performance provided in the Owner's/Flight/Operating Manuals by listing corrected figures against pressure altitude and temperature, it is sometimes valuable to know the **Density Altitude** when the figures given are based upon standard conditions. In effect density altitude is the 'standard conditions' equivalent at any particular altitude and it is found with the aid of the density altitude window provided in most computers (see Fig. 25, page 64).

Imagine you are on an airfield, elevation 500 ft a.s.l., temperature is +30°C and you want to know how this will affect your take-off performance since all figures in your aircraft manual are quoted as for standard conditions.

Method

Using the airspeed window set 500 ft (i.e. half a division) on the Press. Alt. scale against +30°C Air Temperature. In the density altitude window read 2000 ft against the arrow. On this day your aircraft will behave as though it were taking off from an airfield situated 2000 ft above sea level.

Climb Performance

Section 5 of the aircraft manual will give the climbing performance of the aircraft under different temperature conditions. The information will be presented either in the form of a table on the following lines:

WEIGHT (lb)	PRESSURE ALTITUDE (ft)	CLIMBING SPEED (IAS)	RATE OF CLIMB (ft/min)			
			−20°C	0°C	20°C	40°C
1700	s.l.	70	850	780	710	640
	2000	69	750	685	620	555
	4000	68	650	585	520	455
	6000	67	550	490	430	370
	8000	66	455	395	335	275
	10,000	65	355	300	245	190
	12,000	64	260	205	150	95

NOTE: the temperatures are departures from standard.

– or there may be a simple graph.

In the case of small training aircraft with only two seats, the figures will be quoted at one weight. Larger aircraft may also give climbing rates against aircraft weight.

Time, Fuel and Distance to Climb

Another table in Section 5 gives information under the following headings:

WEIGHT (lb)	PRESSURE ALTITUDE (ft)	TEMP (°C)	CLIMB SPEED (IAS)	RATE OF CLIMB (ft/min)	FROM SEA LEVEL		
					TIME (min)	FUEL USED (gallons)	DISTANCE (n.m.)
1700	s.l.	15	70	780	0	0	0
	1000	13	70	700	1	0.2	2
	2000	11	69	685	3	0.4	3
	3000	9	69	635	5	0.7	5
	4000	7	68	585	6	0.9	7

etc., up to 12,000 ft.
NOTE: the temperatures quoted are standard for that altitude.

This table is of particular value during flight planning since it enables the pilot to compute distance flown while climbing, time to any particular cruising level and the amount of fuel used.

Cruising performance

Like climbing performance, flight manual information may be presented in the form of graphs or tables, an example of these being from the Cessna 152 Pilot's Operating Handbook:

PRESSURE ALTITUDE (ft)	RPM	20°C BELOW STANDARD TEMP.			STANDARD TEMPERATURE			20°C ABOVE STAMDARD TEMP.		
		% BHP	KTAS	GPH	% BHP	KTAS	GPH	% BHP	KTAS	GPH
2000	2400	---	---	---	75	101	6.1	70	101	5.7
	2300	71	97	5.7	66	96	5.4	63	95	5.1
	2200	62	92	5.1	59	91	4.8	56	90	4.6
	2100	55	87	4.5	53	86	4.3	51	85	4.2
	2000	49	81	4.1	47	80	3.9	46	79	3.8
4000	2450	---	---	---	75	103	6.1	70	102	5.7
	2400	76	102	6.1	71	101	5.7	67	100	5.4
	2300	67	96	5.4	63	95	5.1	60	95	4.9
	2200	60	91	4.8	56	90	4.6	54	98	4.4
	2100	53	86	4.4	51	85	4.2	49	84	4.0
	2000	48	81	3.9	46	80	3.8	45	78	3.7

etc., up to 12,000 ft.

Range and Endurance Performance

The most usual method of presenting range and endurance information is with simple graphs which show altitude along the vertical edge and range in nautical miles or endurance in hours (as the case may be) along the bottom. Lines are drawn for 75, 65, 55 and 45 per cent power and at intervals the TAS is shown. A 45 min. reserve is built into the figures.

Some graphs are presented under the heading *Best Power Range* and *Best Economy Range*. These relate to the use of lean mixture which can add to the range at the expense of a small reduction in speed for any power setting.

Flying for Maximum Range and Endurance

For an aircraft to provide the **Maximum Range** it must be flown so that:

(*a*)   The airframe is at its best L/D ratio.
(*b*)   The engine is operating at a height where the power required for best L/D speed entails opening the throttle fully.

The advantages of having the most amount of lift for the least amount of drag will, by now, be apparent to the student pilot but condition (*b*) may be less clear. By climbing until the throttle is fully open to achieve the required power the engine can enjoy maximum breathing, the mixture may be leaned and there will be a good TAS for the IAS at that level.

Usually, light aircraft with normally aspirated engines give their maximum range at a reduced cruising speed while flying at 12,000 ft but a more practical power setting would be full throttle at 8000 to 10,000 feet since this would result in a better cruising speed although range would be less.

Maximum Endurance (i.e. maximum time in the air) is obtained at low level and at the lowest power setting that will sustain level flight.

Full Throttle Altitude

When an aircraft is flown at, say, 75 per cent power the level at which it is necessary to open the throttle fully in order to maintain that power is known as the **Full Throttle Altitude**. Most normally aspirated piston engines reach their 75 per cent power full throttle altitude at 8000 ft. From that level power will decrease until, at around 11,000 ft full throttle altitude will correspond to 65 per cent power.

To obtain maximum range or maximum endurance proper use of

the mixture control is essential. This was explained in Exercise 4, page 36, *Flight Briefing for Pilots, Vol. 1*.

Landing Performance

The aircraft manual will list both **Landing Roll** and **Landing Distance**, the latter being the distance to stop after approaching over a 50 ft obstacle (i.e. the 50 ft screen). The information may be conveyed in graph or table form and the variables are:

(1) Pressure altitude
(2) Temperature departure from standard.

The figures assume no wind but there will be a statement advising the percentage decrease for each 9 kt of headwind (or increase for every 2 kt of tailwind). There is usually a 45 per cent ground roll increase when landing on grass since this decreases braking efficiency.

Larger aircraft have tables that vary according to landing weight.

Weight and Balance

In the days when light aircraft were confined to a pilot and one passenger it was almost impossible to overload them or to position the load in a manner likely to place the aircraft outside its centre of gravity limits and hazard safe flight. However, for some years the tendency in light aircraft design has been towards operational flexibility in so far as a pilot can elect to vary the amount of cabin load relative to fuel carried and *vice versa*. It is therefore important that the following terms are understood.

1. Maximum Total Weight Authorized (MTWA). Sometimes known as **Gross Weight**, this is the highest weight at which the aircraft may take off.

2. Maximum Ramp Weight. As the term implies this is the maximum weight to which the aircraft may be loaded. It represents MTWA plus additional fuel for taxying and is mainly applicable to larger aircraft, particularly those powered by gas turbine engines which burn a considerable weight of fuel while taxying and idling on the ground.

3. Maximum Landing Weight. While this does not relate to light single-engine aircraft, some light twins and most of the larger designs are not stressed for landing at MTWA. Consequently, when flight planning in these aircraft it is important not to take on excessive fuel which could, at the end of a short flight, leave the aircraft over-weight for landing, a situation that would require the pilot to remain in the air until sufficient fuel had been consumed to bring the aircraft within its landing weight limits. Maximum landing weight is dictated by structural considerations.

4. Zero Fuel Weight. Larger aircraft intended for cost effective passenger/freight carrying are designed to lower 'g' factors than, for example, trainers. In a large aircraft a considerable proportion of the bending loads on the wing spars is offset by the weight of fuel carried in the wings, consequently as fuel is used these bending loads may actually increase although the total weight of the aircraft has decreased. To cater for this factor **Zero Fuel Weight** represents the maximum weight to which the aircraft may be loaded *exclusive* of fuel. In effect this limits the maximum weight that may be placed within the fuselage. Zero fuel weight is not a limitation that applies to single-engine aircraft or the smaller light twins but it affects more complex designs such as high performance piston-engine twins, turboprops, business jets and, of course, transport aircraft.

5. Empty Weight. This is the weight of the aircraft, its equipment, oil and unusable fuel before passengers, baggage, freight and fuel are loaded.

6. Useful Load. By subtracting Empty Weight from Maximum Total Weight Authorized or, when one is quoted, Maximum Ramp Weight the Useful Load may be found. How it is used (i.e. passengers and baggage *v* fuel) is at the discretion of the pilot provided the aircraft is loaded so that it remains within its **Centre of Gravity Limits** (page 150).

All weights and limitations will be found in Section 6 of the Owner's/Flight/Operating Manual for the aircraft but in so far as single-engine designs are concerned only descriptions 1, 5 and 6 apply.

Baggage stowage and Maximum Weight Allowed

Since many modern light single and twin-engine aircraft are fitted with fuel tanks which, if filled when all seats are occupied could exceed the maximum total weight authorized (MTWA), it is rarely possible to fill the tanks and fly all occupants with their luggage. In most designs payload must be traded for fuel carried, or in other words, range. The consequences of overloading a single engine aircraft are poor take-off and climb performance together with a higher stalling speed and, because of the excess weight, a lower factor of structural safety. Twin and multi-engine aircraft are similarly affected and suffer in addition from a decrease in engine-out performance. This could under certain conditions mean the difference between climbing out safely when an engine fails after take-off and a refusal to gain or even maintain height. The lesson is clear – an overloaded aeroplane is a dangerous one. Most aircraft have a designated baggage area, often provided with a separate door in the side of the fuselage. Baggage compartments are placarded with the **Maximum Authorized Weight** to be carried and this must on no account be exceeded because:

(*a*) The compartment floor will be stressed for no more than the Maximum Authorized Weight.

(*b*) Excess weight in the baggage compartment may place the aircraft outside its Centre of Gravity Limits.

Since baggage compartments are usually situated behind the occupants and therefore some distance from the wing, an out-of-balance situation of this kind will make the aircraft tail-heavy thus moving the centre of gravity nearer the tail surfaces and, in effect, shortening the moment arm through which the tail surface can control the aircraft. A tail-heavy aircraft is more likely to spin if mishandled than a nose-heavy one and the danger of incorrect balance should be understood by pilots particularly since the shortened moment arm also limits the amount of rudder power available to affect spin recovery (Fig. 128).

It is usual to provide tie-down points in the baggage compartment and while on a fine day it may be tempting to disregard these it is always best to assume that at some stage of the flight turbulence may be encountered when loose objects in the back of an aircraft could move and even cause damage.

Calculating Weight and Balance

Ascertaining that an aircraft is not overloaded prior to take-off is in itself not enough to ensure safe flight. The useful load must be positioned correctly otherwise there is a danger that the Centre of Gravity of the aircraft may not coincide with its Centre of Pressure. Obviously it would not be possible to load an aircraft so that the two forces, Lift and Weight, were perfectly aligned; the C of G changes as fuel is used or even when an occupant leans forward, and the centre of

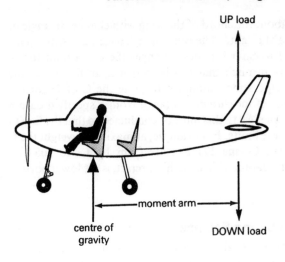

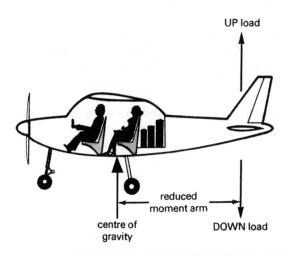

Fig. 128 Movement of the centre of gravity and its effect on the leverage of the tail surfaces.

pressure moves about the chord of the wing with changes in angle of attack (Principles of Flight, "The Aeroplane", page 142). Although a tailplane is fitted to cater for these changes there is a limit to the degree of stability in pitch that can be provided in flight and the Owner's/Flight/Operating Manual will quote the C of G for the aircraft as a range of measurements taken from a specified datum point on the airframe. All flights must be conducted so that the C of G remains within these limits. For example, at maximum weight the C of G limits for the Cessna 177 RG are from 101 in. to 105.8 in. measured aft of the datum which, on this aircraft, is the lower portion of the firewall.

Method of Computing Balance

Method of calculating weight and balance is shown in the manual for the aircraft. Unfortunately the manufacturers have yet to standardize these procedures so the following text is of necessity written in general terms. It is essential that pilots should refer to the manual and be in no doubt as to whether or not the aircraft is in balance before take-off.

The principle of all weight and balance calculations is the simple one illustrated in Fig. 129. Expressed in words it is that the force exerted by a weight on a beam is dependent upon its distance from the pivot point. In aircraft terms the pivot point should be regarded as the required position of the Centre of Gravity. Thus 100 lb (or kg) weight situated 50 in. (or mm) from the pivot point (C of G) will balance 50 lb (or kg) positioned 100 in. (or mm) on the other side of the pivot. By multiplying weight × distance the answer, known as the **Moment**, may be expressed as **pound:inches** or, when metric units are used, **kg:mm**.

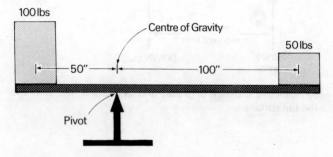

Fig. 129   Weight X Distance = Moment.

Starting with the empty aircraft and its known moment (found previously by weighing, after all equipment has been installed) it is then necessary to compute the moments for all variable load i.e. passengers, baggage, fuel, etc. This is not the daunting task it might at first appear and to make it easier several methods are used. The manual may include a plan of the cabin/baggage area with each seating, baggage or fuel location (known as a **Station**) marked. To save calculating the moment for each station, which would entail multiplying the distance from datum to station by the load in that position, tables are provided listing weights as a scale against which are shown the moments for that station.

For small aircraft it is sometimes the practice to employ a simple graph (Fig.130) showing weight along the vertical edge.

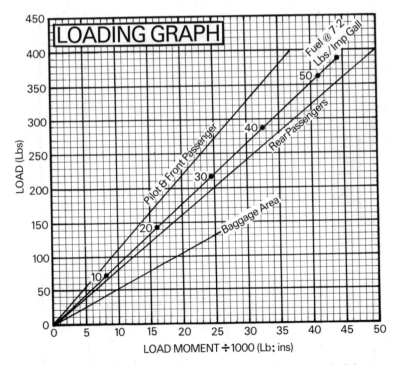

Fig. 130 Typical loading graph of the type included in Section 6 of the aircraft's manual.

LOAD SHEET	Weight (Lbs)	Moment (Lb/ins÷1000)
1. Basic Empty Weight (use data applicable to aircraft)	1800	186
2. Usable Fuel (@ 7·2 Lbs/Imp Gall)	260	30
3. Pilot and Front Passenger	340	31
4. Rear Passengers	318	40
5. Baggage Area	50	9·5
6. TOTAL WEIGHT & MOMENT	2768.	296·5

Fig. 131 Simple load sheet.

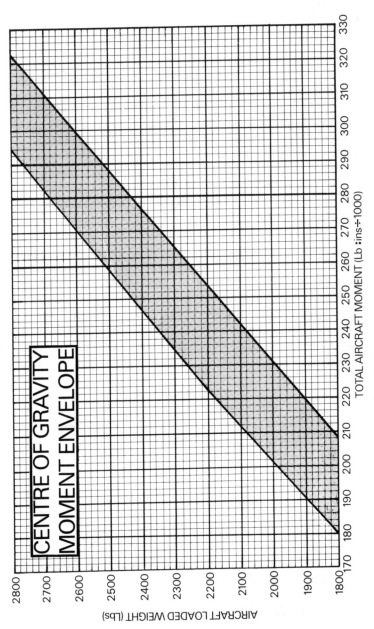

Fig. 132 Aircraft Weight/Moment must fall within the shaded area.

Lines radiate from the, bottom left hand corner, one for each of the following stations:

Pilot and front passenger
Fuel
Rear passengers
Baggage.

In the case of a six-seat aircraft there would be another radial for the rear pair of seats.

As an example imagine that total weight in the two front seats is 350 lb. Trace the 350 lb line across to the 'pilot and front passenger' radial then vertically down to the scale of moments at the bottom of the graph which, in this case shows 32. (*Note.* The moment is actually 32,000 lb:in. but to avoid carrying many noughts throughout the calculations the moments are divided by 1000).

The manual will provide a specimen **Load Sheet** for the aircraft and this will have a line for each station where the weight can vary (e.g. baggage compartment, row of seats, etc.). Weight to be loaded at each station is entered in the 'weight' column and the moment for that station, which has been found using the procedure outlined in the previous paragraph, is inserted in the 'moment' column. When the load sheet has been completed the two columns are then totalled to reveal the total weight of the aircraft and its moment at that weight (Fig. 131). It only remains to relate these two figures to a simple diagram known as the **Centre of Gravity Moment Envelope** (included in the aircraft manual) to determine if the aircraft is within its C of G limits (Fig. 132).

Whenever the aircraft is over weight or outside its C of G range it must be re-loaded before flight.

The various graphs, load sheets and C of G moment envelopes will be found in Section 6 of the aircraft manual.

Questions

1 The highest weight at which an aircraft may take off is known as:
    (a) Maximum Total Weight Authorised,
    (b) Maximum Ramp Weight,
    (c) Certified Take-off Weight.

2 The Owner's/Operating/Flight Manual for larger aircraft usually quotes Zero Fuel Weight. What does this mean?
(a) The legal minimum fuel that may be carried for any flight,
(b) The maximum weight authorised exclusive of fuel,
(c) The maximum weight authorised inclusive of fuel.

3 The Useful Load of an aircraft may be determined by:
(a) Subtracting Empty Weight and weight of fuel from Maximum Authorised Take-off Weight,
(b) Subtracting the maximum weight of fuel from Gross Weight,
(c) Subtracting the Equipped Empty Weight from Maximum Authorised Take-off Weight.

4 It is particularly dangerous to load an aircraft so that its C of G is aft of limits because:
(a) There is an increase in stability and the controls have less affect,
(b) A tail-heavy aircraft is a potential spinning risk and stability is degraded,
(c) Too much weight is concentrated on the mainwheels of the undercarriage.

5 Centre of Gravity limits in the aircraft are quoted in the Owner's/Operating/Flight Manual in the following form:
(a) A total weight,
(b) A number of pound-inches of Kg-mm,
(c) Two figures in inches or mm measured in relation to a particular datum point.

6 A beam is balanced at its centre and a 200 kg weight is then positioned 3 metres from the pivot point. To restore balance what weight must be placed at a distance of one metre on the other side of the pivot?
(a) 600 kg,
(b) 600 kg less half the weight of the beam,
(c) 66.66 kg.

8 Aircraft Performance, Weight and Balance

7 **A Loading Graph is used:**
(a) To calculate airfield take-off performance against aircraft weight,
(b) As a simple means of calculating moments for each passenger, baggage and fuel station,
(c) To ensure that the correct proportion of fuel and payload is loaded into the aircraft.

8 **A Load Sheet is used:**
(a) To calculate total weight of the aircraft,
(b) To calculate total movement,
(c) To calculate total weight and total moment.

9 **An aircraft that normally lifts off at 70 kt takes off with a 7 kt tailwind. This will:**
(a) Reduce the rate of climb,
(b) Reduce the take-off run by 20%,
(c) Increase the take-off run by 20%.

10 **An aircraft that is taking-off from a 'hot and high' airfield will:**
(a) Need a longer run than usual,
(b) Need a shorter run than usual,
(c) Take the same run as usual but lift off at a higher than average indicated airspeed.

12 **The term 'Take-off Distance' when related to a light aircraft means:**
(a) The runway length required for take-off at maximum weight and under ISA, sea-level conditions,
(b) The distance required for the aircraft to reach a height of 50 ft amsl at maximum weight and under ISA, sea-level conditions,
(c) The distance required for the aircraft to reach a height of 50 ft above the runway at maximum weight under ISA, still air, sea-level conditions.

13 **On a hot day, when the temperature is above ISA:**
(a) Take-off distance will increase and rate of climb decrease,
(b) Take-off distance will decrease and rate of climb increase,
(c) Take-off distance and rate of climb will remain unchanged.

14 For any percentage power setting and cruising altitude an increase in temperature above ISA will:
  (a) Result in a decrease in TAS,
  (b) Have no effect on TAS,
  (c) Result in an increase in TAS.

15 When flying for maximum range the aircraft must:
  (a) Fly as low and as slowly as possible,
  (b) Fly as high and as slowly as possible,
  (c) Fly at or near best L/D speed and at an altitude where the throttle is fully open to maintain that speed.

Chapter 9
Aircraft Instruments

Up to the 1920s even large aircraft were equipped with few instruments other than those relating to engine parameters. Then gradually, as the need became apparent for a means of replacing loss of external visual references when cloud, fog or darkness removed from view that all-important pilot's datum, the horizon, various **Flight Instruments** began to appear. Now, even a light trainer is almost invariably fitted with a full **Flight Panel** with the instruments arranged in an internationally agreed order known as the **Basic T** (Fig. 133).

Aircraft instruments fall into three main categories:

1. Pressure-operated
2. Gyroscopic-operated
3. Engine and Ancillary.

The purpose of this chapter is to provide an elementary working knowledge of these instruments, to describe the information provided and to give limitations when these exist.

Pressure-operated instruments

Pressure-operated instruments are used to read aircraft performance in terms of airspeed, height/altitude and rate of climb/descent. The Pressure/Static system was described on page 258 of Chapter 7.

The Airspeed Indicator

This instrument may be calibrated in knots, miles per hour (or kilometres per hour in some countries). It measures the pressure difference between static and dynamic air by employing a diaphragm or a capsule. When the aircraft is stationary both tubes transmit static pressure, but during forward movement, dynamic pressure is built up through the open-ended pitot tube and the diaphragm will distend

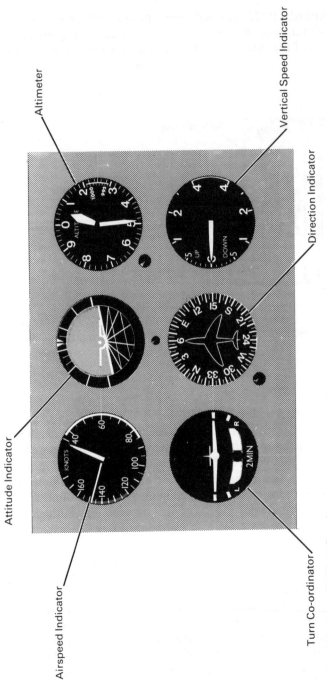

Fig. 133 Basic 'T' Flight Panel.

Altimeter

Vertical Speed Indicator

Direction Indicator

Attitude Indicator

Airspeed Indicator

Turn Co-ordinator

accordingly. In effect the instrument measures the pressure difference between the two samples, the dial being calibrated in units of speed instead of pressure. Fig. 134 illustrates the complete installation.

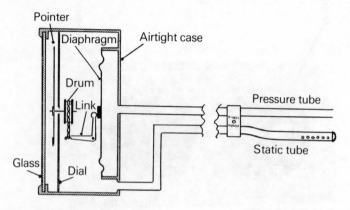

Fig. 134 Simplified drawing of an airspeed indicator. In modern aircraft the pressure and static tubes are contained in a combined Pitot Head or there may be a Pressure Head and separate Static Vents as shown in Fig. 126.

Errors

While there is no lag in the instrument itself, an aeroplane, like any other vehicle, requires time to change from one speed to another so that an alteration in attitude will not produce an immediate change in airspeed. This important fact is sometimes forgotten, resulting in airspeed chasing during instrument flying. The positioning of the pressure head is critical and even in the best installations errors are caused which are due to airflow disturbances. This **Position Error** is taken into account along with **Instrument Error** (inaccuracies in the instrument itself) when converting IAS to RAS. A correction card will be found adjacent to the instrument except when the position and instrument error is small. For practical purposes IAS is often treated as RAS while flying training aircraft.

Of more importance is the effect of varying air density which in turn is dependent upon height (pressure) and temperature. The difference between RAS and TAS resulting from these factors can be very considerable and the following examples will illustrate the importance of calculating TAS for navigational purposes.

Height	Outside Air Temp. (°C)	RAS	TAS
2000 ft	20°	140 kt	146 kt
5000 ft	14°	140 kt	153 kt
10,000 ft	4°	140 kt	165 kt
20,000 ft	−16°	140 kt	192 kt
30,000 ft	−34°	140 kt	230 kt

The Airtour and other computers will calculate a TAS when height and outside air temperature are set against the appropriate scales.

The serious consequences of taking off in Instrument Meteorological Conditions without first removing the pressure head cover are very obvious.

The Altimeter

The purpose of the altimeter is to indicate vertical distance accurately above a pre-set datum which may be airfield level, sea level or a standard datum used by all aircraft flying in the area. In its simplest form the instrument consists of an airtight box which is connected to the static line/vent so that the interior of the altimeter is at the pressure of the surrounding air. An **Aneroid Capsule**, partially evacuated of air, is situated in the box. The capsule is prevented from collapsing by a powerful leaf spring. As height is gained the air pressure in the box decreases allowing the spring to expand the capsule. A linkage system magnifies this movement conveying it to a dial. In reality the instrument is an aneroid barometer, its scale calibrated in units of height instead of pressure. To compensate for the errors induced by expansion and contraction of the mechanism due to temperature changes, a bi-metal strip is incorporated in the linkage which distorts in a corrective sense according to the temperature. The principles of the instrument are illustrated in Fig. 135.

The Sensitive Altimeter

Aircraft are fitted with a development of the simple altimeter just described. The instrument incorporates several capsules and it is manufactured to very high standards. A much greater degree of magnification makes possible a large finger which completes one sweep of the dial every 1000 ft. The dial is numbered from 0 to 9 and

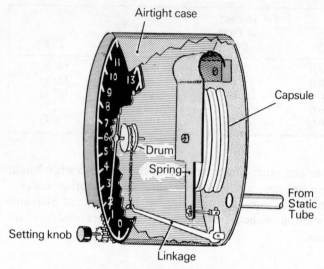

Fig. 135 Simple altimeter. The three-finger type used in most light
aircraft utilise a system of clock-type gears to magnify
expansion and contraction of the capsule and move the
pointers around the instrument dial.

each main division is subdivided into five smaller ones. When reading
the largest pointer each small division represents 20 ft and the main
numbered divisions hundreds of feet. A second smaller finger moves
at one tenth the rate of the 'hundred foot' pointer indicating
thousands of feet, while a third still smaller finger reads in tens of
thousands.

As it is a pressure instrument subject to changes in barometric
pressure, provision is made for re-setting the datum of the altimeter.
This may be accomplished by turning the setting knob until the
instrument indicates zero or airfield level as the case may be.
Alternatively an altimeter setting can be passed over the radio to the
pilot. This is set on a small subsidiary scale on the instrument. When
the altimeter is required to indicate airfield level (i.e. its height above
sea level) the airfield QNH should be requested while a QFE will
cause it to read zero on landing.

The subsidiary scale is calibrated in millibars or, on American
registered aircraft, inches of mercury. The setting knob alters the
reading on the subsidiary scale while at the same time moving the
fingers.

Errors

The instrument is subject to a number of errors and these are:

(*a*) instrument error   (*d*) temperature error
(*b*) pressure error   (*e*) lag error
(*c*) barometric error

In practice the pilot need not concern himself with all of these discrepancies although the following should be understood.

(c) Barometric Error

The instrument is designed to function in accordance with set conditions (ICAN Law) which stipulate a barometric pressure of 1013.2 millibars at mean sea-level and a temperature of plus 15°C. It follows that when, as is so often the case, these average conditions do not exist, the altimeter will fail to give a correct reading. For example, were the pilot to set his altimeter to zero while on the ground when the pressure is 1010 mb, a pressure rise to say 1015 mb would cause the instrument to read 150 ft below zero while a decrease in pressure to 1000 mb would make the altimeter indicate 300 ft above sea level. Furthermore, since barometric pressure varies, sometimes considerably, from one area to another it is not uncommon to fly from a high-pressure region to a low when the altimeter will over-read (or under-read when flying from 'low' to 'high'). Caution must be exercised particularly when flying into lower pressure under instrument flight conditions.

Altimeter setting procedure is explained in Chapter 3, page 120.

(d) Temperature Error

Instrument inaccuracies due to temperature changes are largely compensated for by the bi-metal link, but temperature also affects the atmosphere which provides pressure for the instrument. Under the ICAN standards which form the basis of the sensitive altimeter's calibration, a standard temperature **Lapse Rate** (temperature change per 1000 ft) is assumed. When the m.s.l. temperature departs from 15°C and/or when the lapse rate differs from the ICAN standard atmosphere, air density will likewise alter. Cold air, being heavier than warm, produces greater pressure at any given height and vice versa. For example at 10,000 ft the temperature is assumed to be −5°C under the ICAN Law, but should the outside temperature be −10°C when the reading is taken at that height the resultant higher density will cause the altimeter to read 10,000 ft when the aircraft is at 9800 ft. (Fig. 136) Temperature corrections to height are calculated by

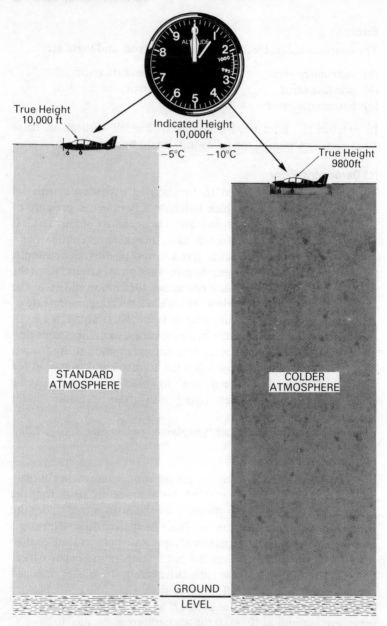

Fig. 136 The effect of temperature on altimeter accuracy. The aircraft on the left is flying under standard atmosphere conditions. That on the right is flying when the temperature is 5 degrees below normal at 10,000 ft. As a result it is 200 ft below the height indicated on the altimeter.

applying the reading from an outside air temperature (**OAT**) thermometer to an Airtour or other computer.

(e) Lag Error
Unlike the airspeed indicator and largely because of the very high degree of magnification involved, altimeter readings will lag, particularly when rapid changes of altitude take place. At rates of climb within the capabilities of most light aircraft, this will not be noticeable, although a prolonged steep descent will cause the altimeter to over-read on levelling out by up to several hundred feet until the instrument has time to settle to the correct reading.

Accuracy
While the instrument is sufficiently sensitive to register changes in height of 20 ft or less its accuracy, due to instrument errors, is not of that order. An average instrument should be within +30 ft to —45 ft at sea-level, the error increasing to ±350 ft at 30,000 ft.

Comparative Checks
Pilots under training will normally fly light aircraft with a single altimeter. Larger, more comprehensively equipped aircraft usually carry two altimeters, one of which may be set to QFE, the other to QNH.

It is essential, prior to departure, to check that both altimeters agree when set to the same QFE/QNH. Likewise, when a single altimeter is set to QNH it should read airfield elevation.

The Vertical-speed Indicator
The VSI measures rate of climb or descent and, during a let-down procedure, assists the pilot to maintain a constant known rate of descent.

The operation of the instrument is in many respects similar to the altimeter in so far as it consists of a capsule inside an airtight case together with suitable magnifying linkage between the capsule and the indicating finger. The VSI is connected to the static line which leads from the static tube or static vent to the altimeter and the airspeed indicator. Inside the instrument the static line is coupled to the capsule so that its internal pressure is the same as that of the surrounding air at whatever height the aircraft may be flying. Air at static pressure is also led into the case of the instrument via a small choke which is finely calibrated so that pressure can change at a constant known rate.

When the aircraft is flown at a steady height, pressure will be the same both within and outside the capsule and the instrument will read zero. Should a climb or descent occur, pressure inside the capsule will change accordingly, whereas the restricted choke will cause the pressure outside the capsule to alter at a slower rate. The capsule will contract (climbing) or expand (descending) until level flight is resumed when the restricted choke will allow the pressure within the instrument case to equal that inside the capsule. It will then return to its normal shape causing the instrument to indicate zero once more.

Because the restricted choke is calibrated to leak at a steady rate, a quick gain or loss of height will produce a correspondingly greater difference in pressure between the inside and outside of the capsule causing it to expand or contract more extensively when a high rate of climb or descent will be indicated. A setting screw is provided for adjusting the capsule so that when static pressure is steady a zero indication is given.

The restricted choke is compensated for changes in air density which occur with height and which would otherwise affect the rate of flow through the orifice. Temperature changes resulting from a climb or descent distort a bi-metal strip to which is attached a small conical projection which alters the size of the choke (Fig. 137).

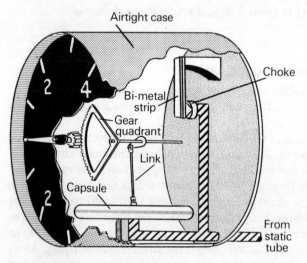

Fig. 137 Vertical Speed Indicator.

Errors

The instrument is affected by position error but this is of little consequence unless the aircraft's ASI is subject to a considerable error under high degrees of acceleration or deceleration. In such cases the VSI will give a false reading.

There is a little lag with the instrument but this is only noticeable when a sudden climb or descent occurs and then the lag is confined to one or two seconds.

Should the static line become blocked owing to an obstruction or ice, the instrument will fail to indicate any changes in height. An **Alternative Static Source** controlled through a small valve, usually located under the instrument panel, allows the VSI, altimeter and ASI to obtain static air supply from within the cabin. The facility is intended for emergency use only since the accuracy of all pressure instruments is affected.

Accuracy

Accuracy of the VSI is within ±200 ft per minute unless the outside air temperature is above 50°C or less than –20°C when errors are in the region of ±300 ft per minute.

Before take-off the pilot should check that the instrument indicates zero.

The Need for Other Instruments

The pressure instruments so far described give an incomplete picture to the pilot when flying without visual reference because their information is confined to:

1. Airspeed (ASI)
2. Height (Altimeter)
3. Rate of height change (VSI).

Certain important factors are missing in the readings provided by the pressure instruments and these are:

1. Heading
2. Attitude in the lateral plane
3. Slip or skid
4. Rate of turn
5. Attitude in the pitching plane.

In view of the relationship between pitch attitude and airspeed, to some extent the last requirement is catered for by the airspeed indicator while requirement 1 (Heading) may be obtained from the magnetic compass. The turn and balance indicator will provide items 2, 3 and 4 and this instrument will be described in detail later. An aircraft fitted with a turn and balance indicator together with an altimeter, airspeed indicator and magnetic compass is said to have a **Limited Panel** and, while instrument flying is practicable in an aeroplane so equipped, demands made on the pilot are somewhat fatiguing.

The vertical-speed indicator does not normally form part of the limited panel but its addition, together with an **Attitude Indicator** and a **Direction Indicator** each of them **Gyroscopic-operated**, brings the flight panel up to the standard illustrated in Fig. 133. The flight instruments are built to very fine limits and the panel is mounted on flexible attachment points to prevent damage from shock or vibration.

Gyroscopic-operated Instruments

Before the gyroscopic-operated instruments are explained in detail, the behaviour of a gyroscope should be understood. When a wheel of any kind is rotated there is a tendency for the material of which it is made to fly outwards because of centrifugal reaction (Fig. 138).

Because the resultant lines of force emanate radially around its circumference, the wheel tends to remain in its plane of rotation and will resist attempts to alter its position. This property is called **Rigidity** and its magnitude depends upon the speed of rotation and the mass of the wheel or other object (this is equally applicable to an airscrew or a child's top). Rigidity in a gyroscope is used to provide attitude information when no natural visual references exist.

The second important property of a gyroscope is called **Precession**. When an attempt is made to disturb its position the gyroscope will resist the force at point of application. A reaction will occur at a point 90° removed from the original force. (Fig. 139).

The 90° shift from point of disturbance to the reaction is always in the direction of rotation. The rate at which a gyro precesses is dependent upon the degree of the disturbing movement, and this principle is employed to provide information about the rate at which an aircraft changes heading.

Using the properties of rigidity and precession, gyroscopes are variously mounted in instruments according to the information required. The gyroscope may be free to move in one plane only or

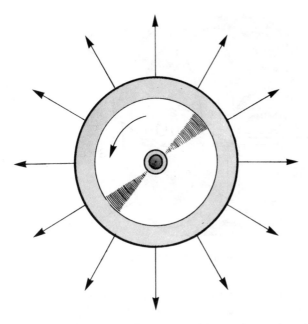

Fig. 138   Rigidity in a spinning wheel. Arrows show outward component of force.

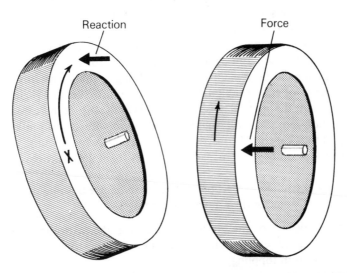

Fig. 139   Precession. Force applied to spinning wheel (right) causes a reaction at 90 degrees in direction of rotation (left drawing).

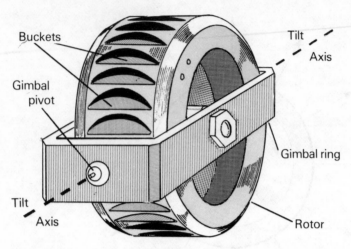

Fig. 140 Air-driven gyroscope mounted in a single gimbal ring.

arranged to allow freedom around all axes. This is accomplished by suspending the gyro in one or more **Gimbal Rings** (Fig. 140).

The gyro illustrated is driven by air which is directed via jets into notches or **Buckets** indented around the outside edge of the gyro. In practice air is drawn out of the instrument by an engine-driven vacuum pump. Replacement air flows into the instrument after passing through a suitable filter and finally discharges through jets, rotating the gyro at some 9000–12,000 RPM according to the type of instrument. Because the jets of air must be directed constantly on the periphery of the gyro and as a result of certain other mechanical difficulties, angular movement of the gimbal rings is limited by stops in instruments of earlier design. Should the aircraft take up a flight attitude outside the limits of these instruments both attitude indicator and direction indicator will **Topple**, i.e. the gyro will cease to hold itself rigid thus causing the instrument to fluctuate violently until it has been re-set. Some modern instruments have complete freedom of movement.

Electrically rotated gyroscopes are in common use but mainly in larger aircraft, some of their advantages being:

1. Simpler design of the gimbal rings allows the instrument to register more extreme flight attitudes before toppling (or freedom from toppling).
2. Higher rotational speeds are possible with the attendant greater rigidity of the gyro.

3. Whereas the efficiency of air-driven gyros decreases at high altitudes because of reduced atmospheric density, electrical gyros are not similarly affected.
4. Electrical instruments, being completely sealed have no contact with fine dust particles and moisture as is the case with air-operated instruments, however well filtered.

When the flight instruments are vacuum operated it has become common practice, even in light trainers. to fit an electric turn and balance indicator as an insurance against complete loss of gyro instruments in the event of vacuum pump failure.

The Turn and Balance Indicator

Function

The Turn and Balance Indicator is used by the pilot to complete turns at a required rate without slip or skid. It is also the primary reference for balanced flight.

When used as a primary instrument on a limited panel lateral level and direction are maintained on this instrument during straight flight.

Principle

The turn and balance indicator is in fact two entirely separate instruments which, in view of their close relationship, are mounted within one casing. The ball indicates slip or skid while the needle shows rate of turn to the left or right.

Turn Indicator

This consists of a gyro which, apart from its own rotation, is free to move in one plane only. The gyro is suspended in a single gimbal ring so that rotation is in the vertical plane. When the complete assembly is turned to the left or right the gyro is made to alter its position as if a force had been applied to one side. Precession causes the gyro to tilt along with its gimbal ring which is restrained by two springs. The gimbal ring ceases to increase its tilt when the force of precession is equalled by the pull of the spring under tension. Should the rate of turn increase, the gyro will precess more powerfully causing the gimbal ring to tilt at a greater angle until spring tension becomes powerful enough to balance the gyro at its steeper angle.

When the turn stops, precession ceases allowing the springs to level the gimbal ring and return the gyro to the vertical position. A pointer is attached to the gimbal indicating rates of turn to the left or right

9 Aircraft Instruments

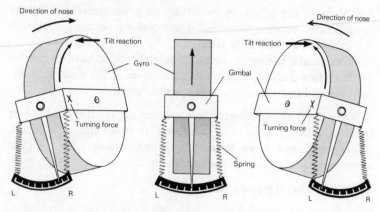

Fig. 141 Principle of the turn indicator (simplified). From left to right: Turning right; straight flight; turning left.

and movement of the assembly is damped by a small piston and cylinder to prevent erratic indications (Fig. 141).

The springs are adjusted so that when the instrument indicates **Rate 1**, the aircraft is changing heading at 3° per second or 180° in one minute. For practical purposes this rate of turn is important since it is used for most instrument procedures. The dial is usually calibrated up to Rate 3 or 4.

Balance Indicator

This takes the form of a ball retained within a transparent tube. In a correctly executed turn the ball will line up with the normal axis of the aircraft so indicating neither slip nor skid.

A turn with insufficient bank or too much rudder pulls the ball towards the outside of the turn, indicating a skid, while insufficient rudder or too much bank cause the ball to move towards the centre of the turn, indicating a slip.

Some balance indicators take the form of a second finger, smaller in size than the turn needle but these are now rare.

Errors

Provided the vacuum supply to the instrument is within the limits laid down (a suction of 4.6 to 5.4 in. of mercury), no errors occur during normal flight and because the gyro is free to pivot in one plane only toppling cannot occur. Reliability is of a high order.

Under conditions of 'g', exaggerated turning indications are given when yaw is present; for instance during the early stages of a loop the needle may indicate a Rate 4 turn and care must be exercised while practising 'Recovery from Unusual Attitudes' not to misunderstand the instrument when the manoeuvre involves abnormal loading.

Before flight the pilot may check the instrument for correct function by watching its behaviour during turns on the ground. The turn needle should indicate a turn in the correct direction while a skid in the opposite direction will be shown by the ball. When the aircraft is stationary and standing level, both needle and ball should be central.

Turn Co-ordinators

In light aircraft there has been a move to replace the traditional needle and ball presentation (or in some instruments, two needle presentation) turn and balance indicator with a **Turn Co-ordinator**. These instruments display yaw/turn rate information in the form of an aircraft symbol which banks against a scale marked on the dial at each wing tip. A ball-type balance indicator is built into the lower section of the instrument face.

The axis of the operating gyro is tilted slightly to improve instrument response to roll/yaw. While certain advantages are claimed for these instruments it should be understood that, although they look like small attitude indicators, no pitch information is provided. Indeed, some flying schools add a small notice to this effect on the glass face. The instrument is intended to show pictorially when the correct bank has been applied for a turn. However it will also show bank when a flat, wings-level yaw is enacted and to this extent the instrument can be misleading.

The Attitude Indicator

Function

The Attitude Indicator is essentially a gyroscopic-operated instrument displaying indication of pitch and roll in place of the horizon when no natural features can be seen. The information is shown pictorially by employing a small aircraft symbol together with a horizon screen. Additionally a pointer denotes angle of bank so that when the relationship between bank and rate of turn is known, turns may be executed on this instrument alone, always remembering that the angle of bank for a given rate of turn will depend upon airspeed.

Because of the very wide range of information provided by the attitude indicator, this instrument is of great importance to the pilot.

Principle

A small aircraft symbol is suspended within the face of the instrument representing the real aircraft as seen from behind.

Because it is fastened to the instrument case the symbol will move with the aircraft. A gyro-controlled screen, depicting a blue sky, the horizon and the ground (in a dark colour) is positioned behind the aircraft symbol and when the instrument is in operation a pitch or roll will cause the small aircraft in the instrument to move accordingly in relation to the horizon which is held rigidly by its gyro so that it conforms to the real horizon. This description is in fact an over-simplification since, although correct when applied to movement in the rolling plane, it is necessary to introduce a reversal mechanism for pitch indications; otherwise a dive would be shown during a climb and vice versa.

Presentations vary slightly from one instrument model to another but usually there is an angle of bank scale around the top of the dial supplemented by angle lines drawn on the 'ground' section of the horizon screen. These also give a 3D effect.

Mechanically the instrument utilizes a gyroscope rotating on a vertical axis. The gyroscope spins within a case which acts as an inner gimbal. The rotor case/inner gimbal is suspended in an outer gimbal, and movement between the two actuates the horizon screen in the pitching plane via a simple reversal mechanism which is shown in Fig. 142. The outer gimbal is free to tilt in the rolling plane so that the aircraft may bank while the gyro (and horizon screen) remains level. (Fig. 143).

There are various artificial horizon models in use and, while earlier instruments have toppling limits of 110° in bank and 60° in pitch, later designs allow complete freedom in roll and 85° in pitch. Others have complete freedom in both axes.

When toppling occurs the instrument will re-set automatically although some ten to fifteen minutes is required for full recovery. During the re-setting period the horizon screen will move random fashion across the face of the instrument, the oscillations becoming less as the automatic re-setting device brings the gyro under control. Automatic erection of the gyro is accomplished by a valve suspended below the rotor called the **Pendulous Unit**. The pendulous unit also

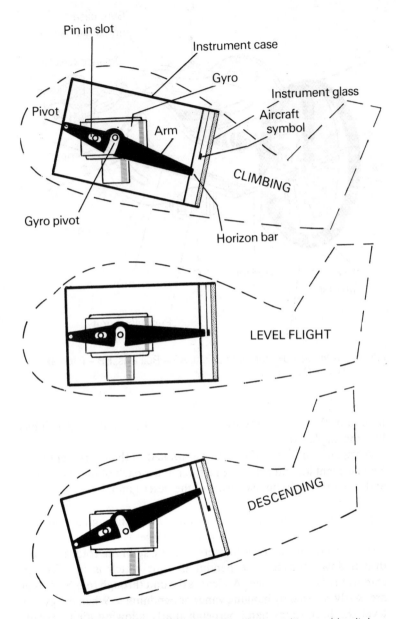

Fig. 142 Attitude indicator. Reversal mechanism used to provide pitch
indications in the correct sense.

9 Aircraft Instruments

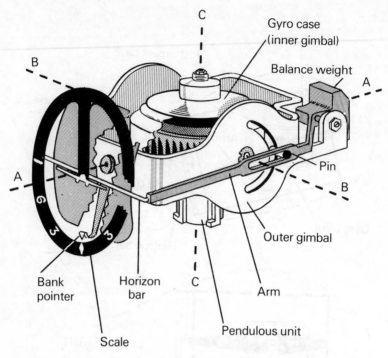

Fig. 143 Attitude indicator mechanism: AA – Roll axis; BB – Pitch axis;
CC – Gyro axis.

maintains the rotor in its correct position when for any reason it has
become displaced.

A vacuum of 3½ in. of mercury is applied to the instrument case.
Replacement air enters through a filter at the back of the instrument
and is led through the rear pivot of the outer gimbal.

A passage in the outer gimbal transmits the air through the pivot of
the rotor case/inner gimbal where it discharges via two jets which are
directed towards the rotor buckets. Having spun the rotor the air is
drawn down through the pendulous unit which is in effect an
extension of the rotor case. Air leaves the unit through four holes that
are partly covered by hanging vanes or pendulums. When the gyro is
level the four vanes hang perpendicularly allowing air to escape
evenly through the four holes (Fig. 144). Should the rotor take up a
permanent tilt in one plane, two of the vanes will be displaced so that
one hole is fully open while that opposite is covered by its vane

306

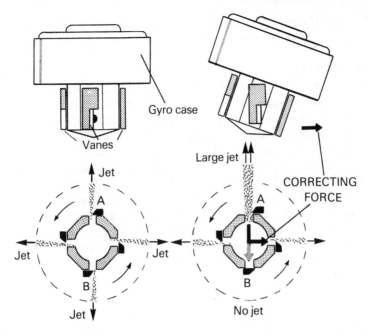

Fig. 144   Pendulous unit. Illustrations on the left (top and bottom) show
position of vanes who the gyro is level. When tilted (right) vane
A opens while B is closed causing a reaction in the direction of
the shaded arrow (i.e. towards B). 90° effect corrects the gyro
returning it to the level position.

causing one of the four air jets to cease and its opposite number to
increase. The air jets will now be out of balance and the rotor will
precess in a corrective sense.

The action described is limited in precessing power requiring some
appreciable time to effect a correction so that displacement of the
vanes during turbulent flight will not cause them rotor to shift its axis.
Air from the pendulous unit is extracted from the instrument case by
an engine-driven vacuum pump.

Errors

In flight, accelerations in the fore-and-aft and lateral planes will
displace the pendulous vanes so producing a 90° precessing effect on
the gyro axis when the horizon screen will fail to correspond with the
true horizon. Furthermore the pendulous unit is suspended below the
tilting axis of the gyro so that any tendency for the assembly to swing

under the influence of acceleration or deceleration will cause the gyro
to precess through 90° and as a result to give an incorrect indication.

(a) Acceleration

While the effect is of less importance in light aircraft or those of
moderate performance, serious displacement of the horizon screen
can occur during take-off with a high-speed aircraft capable of more
vivid acceleration. The combined effect of vane displacement and
pendulous unit inertia during take-off will cause the instrument to
indicate a climbing turn to the right (or, in the case of electric
instruments, a climbing turn to the left) and while the indication is
small an attempt to correct would result in a descending turn in the
opposite direction. This error would be most likely to occur with a
high-performance aircraft when it would be particularly dangerous.
It can be overcome by monitoring the attitude indicator with the turn
and balance indicator and ASI during the early stages of the take-off
and climb away.

(b) Deceleration

The effects of deceleration are the reverse to acceleration causing the
instrument to show a gentle descending turn to the left, although it is
emphasized that such indications will only occur when a high-speed
aircraft reduces airspeed over a lengthy period.

(c) Turning Errors

During a turn, centrifugal reaction will displace the two pendulous
vanes which pivot on a fore-and-aft axis causing the rotor to precess
through 90° in the rolling plane. Additionally the pendulous unit will
be subjected to centrifugal reaction causing rotor precession in the
pitching plane.

   The combined effect of both displacements during a prolonged
turn results in errors which increase until the aircraft has changed
heading through 180°. Continuation of the turn beyond 180° has the
effect of progressively cancelling these errors until 360° is completed
when accurate indications will be given again. These inaccuracies,
which are not of great magnitude, show themselves as incorrect nose
up or down indications coupled with too large or too small a degree of
bank, according to the direction of turn and the number of degrees
change of heading. The errors are partly compensated for by
arranging the gyro to run at a slight angle so that during a rate 1 turn
errors are very small, increasing slightly when heading changes are
made above or below that rate.

Full rotor speed is reached after the correct suction has been applied to the instrument for four minutes although it will operate satisfactorily within half that time.

While on the ground, the attitude indicator may be checked for serviceability by noting that the horizon line is steady and in the correct position relative to the aircraft symbol. Turns during taxying must not disturb the horizon screen. The glass should be free from cracks.

In flight the instrument must indicate immediately any attitude changes in the pitching or rolling planes.

The Direction Indicator

Function

The direction indicator enables the pilot to fly on any heading and to execute turns without acceleration or other errors associated with the magnetic compass. Furthermore the instrument gives an immediate indication when there is a change in direction. Unlike the compass it has no means of seeking magnetic north so that it must be synchronized with the aircraft's compass before use.

Principle

A gyro mounted so that it rotates on a horizontal axis is attached to an inner gimbal ring. This assembly is suspended in an outer gimbal which allows the gyro and its inner ring to rotate through 360° in the horizontal plane. Air is fed to the rotor buckets via a double jet, its purpose being to make the gyro maintain its horizontal axis should a tilt occur over a period of time. Geared to the outer gimbal is a compass scale usually graduated at 5° intervals although some instruments indicate every degree (Fig. 145).

The gyro with its inner and outer gimbal rings is suspended in an airtight case incorporating a glass front which allows the scale to be seen against an aircraft etched in plan form, its nose acting as a lubber line.

Although an aircraft is banked during turns, the gyro is free to maintain its upright position because of the inner gimbal in one plane and its axis in the other. Instruments of old design had toppling limits of 55° in pitch and bank. These are now increased to 85° and some modern direction indicators have no toppling limits.

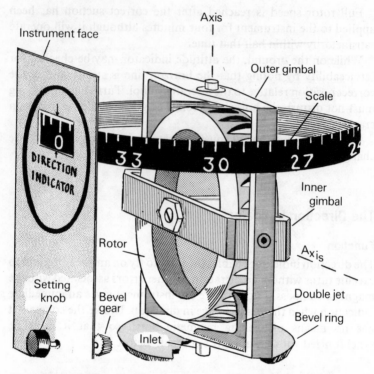

Fig. 145 Direction indicator. For simplicity an old type of instrument is shown in this drawing. Modern DIs have a vertical, rotating dial geared to the vertically-mounted gyro.

The front of the instrument carries a setting knob which is normally disengaged from the assembly within. When the setting knob is pressed in, (a) the inner gimbal ring is moved into the upright position and locked preventing the gyro from tilting in relation to the aircraft, and (b) a bevel gear engages with a bevel ring so that rotation of the setting knob causes the complete assembly to revolve allowing the pilot to synchronize the DI with the magnetic compass. The knob is spring loaded in the 'out' position.

Errors

Over a period of time the DI will fail to agree with the reading shown on the magnetic compass, the difference becoming larger as time advances. This drift from the correct setting emanates from three sources, **Mechanical Drift, Apparent Drift** and **Transport Error**.

(a) Mechanical Drift

Although the DI is manufactured to high standards, bearing friction, imperfect balance of the gimbal rings and severe turbulence in flight will cause the gyro to precess. Furthermore the erecting action of the double jet supplying the rotor tends to displace the outer gimbal ring so that up to 4° drift may occur in a 15-minute period.

(b) Apparent Drift

Apparent Drift is non-existent at the equator. In the extreme case of an aircraft flying over the North or South Pole, the scale of the DI will remain rigid while the earth rotates at its rate of 15°/hour and the DI will appear to have wandered off setting by that amount. As an aircraft flies towards the equator, apparent drift decreases accordingly and in proportion to the latitude. The instrument is compensated by employing a **Drift Nut** which applies a precessing force to the inner gimbal ring so balancing apparent drift. The drift nut completely balances apparent drift at one latitude only and when an aircraft moves to another base some distance north or south of original, the instrument should be adjusted. This is of particular importance when a change is made from the northern to the southern hemisphere or vice versa since apparent drift is the reverse in each case.

(c) Transport Error

Yet another error results from convergence of the meridians towards the poles. In northerly or southerly latitudes an aircraft flying in a straight line on an easterly or westerly heading would cut successive meridians at different angles because they are not parallel. In consequence there will be a changing relationship between magnetic north as indicated on the compass and DI reading which remains fixed in space. This is known as **Transport Error**; it is at its maximum at the poles and does not exist at the equator where, for practical purposes, the meridians run parallel to one another.

The mechanical limitations imposed by the inner and outer gimbal system cause the scale to rotate past the lubber line at an uneven rate when the aircraft is banked, but as the wings become level the gimbal system aligns itself and the correct heading will be indicated. While this **Turning Error** is noticeable at fairly steep angles of bank, it is of little impsortance in instrument flying when turns are limited to a Rate 1 or less.

Vacuum supply at 4–5 in. of mercury is required by the instrument for two minutes before use. The pilot may check the direction indicator for serviceability by seeing that an immediate change in heading is indicated during turns on the ground. There should be a resistance felt when the caging knob is engaged and turned.

Synchronizing the Direction Indicator with the Compass

Great care must be exercised while synchronizing the DI with the magnetic compass; it is essential that the wings are level and the airspeed constant during the process. So that it indicates magnetic heading allowance should be made for deviation when setting the DI against the magnetic compass.

If for any reason the instrument topples, it should be caged and re-set with the magnetic compass.

Because of mechanical and apparent drift the DI must be checked against the magnetic compass every fifteen minutes or less and if necessary re-set.

Heading Indicators

Various improved compass systems are in widespread use and although some of these are confined to larger aircraft simpler versions may be seen in light, single-engine tourers.

In its simplest form the direction indicator previously described is replaced by a similar looking instrument which incorporates a north-seeking device known as a **Flux Gate**. The flux gate, which is a magnetic needle combined with a switching mechanism, is arranged to align the gyro unit with magnetic north, thus eliminating the need for regular resetting of the DI with the magnetic compass. Such a device is often known as a **Slaved Gyro** and the instrument is called a **Gyro-Magnetic Compass**. A further development of the gyro-magnetic compass is the **Horizontal Situation Indicator** which incorporates steering command information from the radio navigational equipment. Student pilots are unlikely to encounter either instrument while under training for the Private Pilot's Licence.

Engine and Related Instruments

In addition to the flight instruments previously described are those provided to allow proper management of the engine. It is usual to group these together in one area of the instrument panel.

In light aircraft many of the engine instruments are supplied by the motor industry and a typical complement of engine readouts to be found in a light trainer would be as follows:

Engine Speed Indicator

Sometimes known as a **Tachometer**, this instrument reads engine RPM. It is usually calibrated at intervals of 100 RPM and the dial will have a red line at the maximum permitted engine speed. Some engines have a small range of RPM within which vibration can be harmful and, in these cases, the manufacturers recommend that such power settings must be avoided. The engine speed range to be avoided will be marked with a coloured arc. The normal range of engine speeds is often marked with a green arc.

Many RPM indicators incorporate a small window through which mileometer-type numbers indicate total engine hours, a useful check on servicing intervals.

Oil Pressure Gauge

This is similar to a car-type oil pressure gauge except that it is marked with a minimum idling pressure (red line), a normal operating sector (green) and a maximum pressure red line.

Oil Temperature Gauge

Within the engine oil system is an electrical resistance-type temperature sensor which is coupled to a temperature gauge on the instrument panel.

Like most modern engine instruments the oil pressure gauge is usually marked with a green arc for normal operating temperatures and a red line to denote maximum. Ideally, the engine should not be run at more than 1000–1200 RPM until the oil temperature is in the green sector. However, on a cold day it is sometimes not possible to achieve this temperature without idling for a long time and the aircraft manual usually advises a minimum warm-up period before opening the throttle for the power checks.

Cylinder Head Temperature Gauge

In many respects this instrument is the equivalent of a car water (cooling) temperature gauge. Most aero engines are air-cooled and, since there is no water, temperature is measured at the cylinder heads.

A cylinder head temperature gauge is not always fitted in a light trainer; it is usually regarded by the manufacturers as an optional extra but like other engine instruments it is colour coded to assist the pilot in instant recognition of what is normal and what is abnormal.

Fuel Contents Gauge

Light aeroplane fuel gauges are similar to those fitted in most cars. In the tank, a float mounted on a hinged arm adopts the level of fuel at the time. As the float arm takes up its position according to the amount of fuel in the tank it alters a rheostat (variable electric resistance) which adjusts readings on the pilot's fuel gauge by altering the balance between two electric currents.

The instrument is based upon a potentiometer – a form of electric circuit which, because it measures differences between currents is therefore little affected by the state of the battery unless it is so undercharged that it cannot work the instrument.

When more than one tank is fitted each will have its own fuel gauge although in some larger aircraft a single gauge reads the contents of fuel in the tank selected. Alternatively, there may be a single gauge with a tank selector switch.

Rheostat/potentiometer-based fuel measuring systems vary from the accurate to the grossly misleading. Furthermore, today's accurate gauge could prove untrustworthy tomorrow. Therefore, before any flight the pilot must visually check fuel contents by looking into the tank through the filler cap.

Larger aircraft are fitted with **Pacitor** fuel measuring systems which have no moving parts other than the finger of the instrument itself. Provided the aircraft is flying with no more than 15° of pitch or roll these instruments will give an accurate reading.

Fuel Pressure Gauge

When the aircraft is of low-wing design an electric fuel pump will be fitted as a back-up in case of mechanical pump failure. In such cases a fuel pressure gauge is provided and when the electric pump is turned on prior to engine starting it will indicate that fuel pressure exists, thus confirming that the carburettor is being supplied.

In the event of engine failure the fuel pressure gauge should form part of the 'cause of failure' checks. Lack of fuel pressure could indicate malfunction of the engine-driven fuel pump, a situation that can be rectified by turning on the electric fuel pump.

Instruments fitted in more advanced aircraft

The engine instruments so far described will be found in most light trainers. Aircraft intended for touring may also have the following instruments:

Exhaust Gas Temperature Gauge

To assist the pilot in setting the most economical mixture for best fuel economy an EGT Gauge is often fitted. It functions by measuring the temperature of the exhaust gas, there being a direct relationship between this and fuel/air ratio. A weak mixture burns hot while rich mixtures burn at lower temperatures.

Manifold Pressure Gauge

When a constant speed propeller is fitted movement of the throttle throughout most of its range does not affect engine RPM which are controlled on the propeller lever (see Chapter 6, page 238).

To provide the pilot with a suitable means of setting the throttle a manifold pressure gauge is provided. This measures the pressure within the induction manifold and the instrument is calibrated in inches of mercury.

Fuel Flow Indicator

Fuel injected engines are provided with fuel flow indicators which, as the name implies, measure the flow of fuel in gallons per hour. Since most light aero engines are of American manufacture these instruments are calibrated in US gallons.

Fuel burns for normal cruising are usually marked with a green sector.

When the aircraft has a constant speed propeller and the engine is fuel injected it is the practice to house both the manifold pressure gauge and the fuel flow indicator within one case.

Ancillary Instruments

Ammeter/Voltmeter

Most light aircraft are fitted with an Ammeter which indicates current flow (in amperes) from the alternator to the battery or from the battery to such electric services as the radio or pitot heat when the alternator has failed or current demand exceeds the output of the alternator.

Engine starting demands a heavy current and immediately after the engine is running a high charge will be indicated as the alternator charges the battery. Gradually the charging rate will decrease as the battery is replenished.

Outside Air Temperature Gauge

So that the pilot may make both airspeed and altimeter corrections in flight, and be aware of icing risks, an OAT gauge is fitted, usually high up on the windscreen behind the temperature probe which measures the temperature of the outside air. An alternative type of OAT gauge is mounted on the instrument panel, some distance from the probe and linked to it electrically.

Although this instrument is regarded by most manufacturers as an optional extra it is nevertheless important for safe operation of the aircraft.

Vacuum Gauge

To operate the attitude indicator and the direction indicator vacuum is supplied by an engine-driven pump.

Satisfactory performance of these instruments depends, among other factors, on their being supplied with vacuum sufficient to spin the gyros at the correct speed. In most light trainers the vacuum system operates in the range of 4.6 to 5.4 in. of mercury and a small vacuum gauge is provided on the instrument panel so that the pilot can check that this is being delivered.

Usually normal suction is marked on the vacuum gauge by a green sector and the instrument should be checked during the engine run-up. Low vacuum could be caused by a blocked filter or an unserviceable vacuum pump. In either case readings of the gyro-operated instruments could be suspect.

Flap and Trim Position Indicators

The simplest method of indicating the amount of flap depression is by marking the flaps with suitable degree positions. While some light trainers use this method it is more usual to provide a small **Flap Position Indicator**. Ideally this should be adjacent, to the flap selector.

Trim position may be indicated on the trim wheel itself or with an elevator Trim Tab Position Indicator. This too is usually adjacent to its control, in this case the trim wheel.

Hydraulic Pressure Gauge

This instrument is only fitted in aircraft of more advanced design that have an hydraulic system for flap and/or undercarriage operation. Its purpose is to enable the pilot to ensure that the hydraulic system is being supplied with sufficient pressure to operate its dependent services.

Warning Lights

It is common practice to fit warning lights in aircraft. These draw attention to such matters as low oil pressure, alternator failure, electric fuel pump on, flaps not fully retracted, low fuel pressure, etc.

Usually the various warning/advisory lights are grouped together. In a light aircraft there will be a strip of such lights along with a 'Press to Test' button for checking that all bulbs are functioning. A more complex aircraft will have a large cluster of these lights. Such an arrangement is known as an **Annunciator Panel**.

Fixed Notices

On the instrument panel and elsewhere in the aircraft are a number of fixed notices giving limitations (weights, speeds and such matters that may affect the safe operation of the aircraft). All notices, which are often known as **Placards**, are listed in the aircraft manual.

Questions

1 Blockage of the static tube or vent will affect the following instruments:
   (a) ASI, Altimeter and VS1,
   (b) ASI, Altimeter and Turn Indicator,
   (c) Altimeter, VSI and Turn Indicator.

2 There is an appreciable elapse of time before the ASI settles to a new airspeed. Why is this?
   (a) Lag in the instrument,
   (b) Inertia of the aeroplane,
   (c) Position error.

3 **The difference between IAS and RAS is caused by:**
   (a) Position error,
   (b) Instrument error,
   (c) Position and Instrument error.

4 **A low air density due to altitude and high temperature will have this effect:**
   (a) Give an increased TAS for any RAS,
   (b) Give a decreased TAS for any RAS,
   (c) Give a TAS that is less than RAS.

5 **The altimeter obtains its pressure sample from:**
   (a) The static line or static vent,
   (b) The pressure line,
   (c) A venturi tube.

6 **A sensitive altimeter at low levels is accurate to within limits of:**
   (a) + or – 20 ft,
   (b) + or – 350 ft,
   (c) +30 or – 45 ft.

7 **What is the standard ICAN Barometric pressure?**
   (a) 1013 mb at mean sea-level and 0°C,
   (b) 1013.2 mb at mean sea-level and at a temperature of +15°C,
   (c) 1013.2 mb at mean sea-level and +10°C.

8 **You are flying into an area of high pressure. The altimeter will:**
   (a) Under read,
   (b) Over read,
   (c) Be unaffected.

9 **For the purpose of instrument flying high rates of descent must be avoided because:**
   (a) The gyro instruments may topple,
   (b) The aircraft may exceed its V_{ne},
   (c) The altimeter will lag seriously under conditions of high rates of vertical change.

10 You take-off from an airfield with an elevation of 150 ft amsl having set the Regional QNH of 1012 mb. The destination is a non-radio airfield situated on a 600 ft hill within the same altimeter setting region. What QFE setting should be used for the landing?
   (a) 997 mb,
   (b) 992 mb,
   (c) 1032 mb.

11 For the purpose of calibrating an altimeter a temperature of –5°C is assumed at 10,000 ft. What effect will a lower temperature have on the aircraft's true height when the instrument reads 10,000 ft?
   (a) It will be above 10,000 ft,
   (b) It will be below 10,000 ft,
   (c) No effect because temperature changes are compensated by a bi-metal link.

12 Within a temperature range of +50°C and –20°C the vertical speed indicator is accurate to within limits of:
   (a) ±200 ft/min,
   (b) ± 30 ft/min,
   (c) ±100 ft/min.

13 Viewed face-on a gyroscope is rotating in a clockwise direction. Where should you apply a force to make it twist to the left (i.e. the right-hand rim will move away from you)?
   (a) Against the top face of the rotating gyro,
   (b) Against the right face of the rotating gyro,
   (c) Against the right edge of the gyro, inwards towards the centre.

14 The following instruments on the flight panel are gyro operated:
   (a) Attitude indicator, direction indicator and turn and balance indicator,
   (b) Attitude indicator, direction indicator and turn indicator,
   (c) Attitude indicator, direction indicator and balance indicator.

15 Why do some gyro instruments topple when the aircraft is placed in an extreme attitude?
   (a) Because the air supply to the gyro is discontinued,
   (b) Because the Pendulous Unit is displaced beyond its operating range,
   (c) Because the Gimbals come up against their limiting stops.

16 Which gyro operated instrument may be relied upon to give factual information during the recovery from a spin (assuming the instruments fitted have toppling limits)?
   (a) The turn indicator,
   (b) The slip indicator,
   (c) The attitude indicator.

17 Before take-off the direction indicator must be synchronized with the magnetic compass. Why is this?
   (a) Because the gyro may not have reached its correct operating speed,
   (b) To allow for local magnetic variation,
   (c) Because a direction indicator has no means of seeking Magnetic North.

18 At regular intervals during flight the direction indicator must be re-set against the magnetic compass. Why is this?
   (a) Because of mechanical drift,
   (b) Because of mechanical drift, apparent drift and transport error.
   (c) Because of turbulence.

19 What feature is incorporated in the direction indicator for the purpose of correcting the effects of apparent drift?
   (a) A spring attached to one of the gimbals,
   (b) An adjustable drift nut attached to one of the gimbals,
   (c) A pendulous unit.

20 What is the principal advantage of a gyro-magnetic compass over a direction indicator?
   (a) There are no toppling limits,
   (b) It is suitable for Polar Navigation,
   (c) It incorporates its own north-seeking device and an automatic synchronizing system.

21 In a vacuum operated attitude indicator, automatic erection of the gyro is performed by:
   (a) The pendulous unit,
   (b) A caging device,
   (c) A counter-weight on the horizon bar.

22 You have just taken-off in a fast aircraft fitted with a vacuum
    operated attitude indicator. While climbing straight ahead the
    instrument may for a short while indicate:
    (a) A high nose-up attitude.
    (b) A climbing turn to the left,
    (c) A climbing turn to the right.

23 How is vacuum provided for the gyro operated instruments?
    (a) By the static tube,
    (b) By an engine-driven pump or a venturi tube,
    (c) By the static vent.

24 What is the meaning of the term 'dip' as applied to the magnetic
    compass?
    (a) The tendency for the magnet system to tilt during turns,
    (b) The residual deviation present after a compass swing,
    (c) The tendency for the magnet system to be pulled down
         towards the Earth's magnetic field.

25 In the Northern Hemisphere what effect will acceleration and
    deceleration have on an easterly or westerly compass reading?
    (a) An apparent turn to South and North respectively,
    (b) An apparent turn to North and South respectively,
    (c) No change.

26 While on a northerly heading within the Southern Hemisphere the
    aircraft is flown left wing low. What effect will this have on the
    compass card?
    (a) The south-seeking edge of the compass card will swing
         towards the lower wing,
    (b) The north-seeking edge of the compass card will swing
         towards the lower wing,
    (c) No change.

27 When using a magnetic compass in the Northern Hemisphere what is the correct technique for turning onto N, S, E and W?
   (a) Roll out of the turn 25°-30° before reaching North, 25-30° after reaching South and 5°-10° before reaching East or West.
   (b) Roll out of the turn 25°-30° after reaching North, 25°-30° before reaching South and 5°-10° before reaching East or West,
   (c) Roll out of the turn 25°-30° before reaching North, 25°-30° before reaching South, 5°-10° before reaching East and 5°-10° after reaching West.

The following questions relate to the use of instruments.

28 Although the required heading is being maintained the ball of the balance indicator is over to the left. What correction is required to resume balanced flight?
   (a) Left rudder until the ball centres,
   (b) Left rudder until the ball centres together with right aileron to level the attitude indicator,
   (c) Right rudder until the ball centres together with right aileron to level the attitude indicator.

29 While practising instrument flying you inadvertently place yourself in a position where the instruments indicate a balanced rate 4 turn to the left coupled with a rapid loss of height and a rapid increase in airspeed. What is the aircraft doing?
   (a) Spinning to the left,
   (b) In a spiral dive to the left,
   (c) A steep turn to the left with the pressure tube blocked.

30 While climbing at full throttle on instruments you note that although the airspeed is correct the rate of climb is considerably below normal. What action should you take?
   (a) Adopt a higher nose attitude on the attitude indicator and re-trim,
   (b) Lower the nose slightly and increase the climbing speed by approximately 5 kt,
   (c) Check for carburettor icing.

31 The attitude indicator has become unserviceable during instrument flying. The aircraft enters a steep dive with the airspeed increasing rapidly. During the recovery what indication from the remaining instruments will tell you when the nose is on or near the horizon?
   (a) When the ASI returns to normal cruising speed,
   (b) When the ASI stops increasing and the altimeter stabilizes,
   (c) At the point where the ASI stops increasing and begins to move towards the original cruising speed.

32 What angle of bank should you adopt on the attitude indicator for a rate 1 turn while flying at an IAS of 130 kt?
   (a) 15°,
   (b) 18°,
   (c) 20°.

33 The direction indicator has failed and it is necessary to continue instrument flight using the magnetic compass as the sole heading indicator. When making timed rate 1 turns through a required number of degrees a pilot should:
   (a) Allow 3° per second and start timing after the correct angle of bank has been attained.
   (b) Allow 3° per second. Start timing when rolling into the turn and roll out after the required number of seconds have elapsed.
   (c) Allow 3° per second. Start timing when rolling into the turn and aim to have the wings level again after the required number of seconds have elapsed.

34 While flying on instruments in deteriorating weather the aircraft enters turbulence and assumes an unusual attitude. The instruments indicate an unbalanced, rate 4 turn, a rapid height loss and a low, fluctuating airspeed:
   (a) The aircraft has stalled,
   (b) The aircraft is responding to turbulence,
   (c) A spin has occured.

Chapter 10

Introduction to Radiotelephony

With the increase in air traffic and expansion of controlled airspace almost all aircraft are of necessity fitted with a VHF radio transmitter/receiver **(Transceiver)**.

At certain airports it is not unusual to find student pilots in the circuit with large commercial aircraft. Safe conduct of such mixed traffic is made possible by a common controlling authority in contact with each aircraft through VHF radio equipment.

At first hearing the student may consider radio messages to and from aircraft to be terse verbal shorthand and in many ways this is an apt description, the transmissions being composed of figures, letters, abbreviations and code words. In fact letters and figures are frequently used and because misunderstandings may result from foreign pronunciations, static or other interference with reception, an international phonetic alphabet has been devised to avoid confusion:

A	Alpha	N	November
B	Bravo	O	Oscar
C	Charlie	P	Papa
D	Delta	Q	Quebec
E	Echo	R	Romeo
F	Foxtrot	S	Sierra
G	Golf	T	Tango
H	Hotel	U	Uniform
I	India	V	Victor
J	Juliett	W	Whisky
K	Kilo	X	X-ray
L	Lima	Y	Yankee
M	Mike	Z	Zulu

Similarly a pronunciation for numerals is prescribed:

0	Zero	5	Fife
1	Wun	6	Six
2	Too	7	Sev-en
3	Tree	8	Ait
4	Fow-er	9	Nin-er

Tens, hundreds, etc., are spelt out, e.g. 10 is 'one zero,' 650 is 'six fife zero,' 25,000 is 'two fife tousand'.

Establishing Communication

When an aircraft has established contact using the five letter registration as the callsign (or in some cases with foreign aircraft, numbers or both) the first and last two letters of the callsign may be used as an abbreviation when this has first been initiated by ATC.

Before transmitting it is essential to listen carefully to ensure that no other transmissions are being made. To transmit over an existing message is to create unnecessary noise and confusion to the inconvenience of other aircraft operating on that frequency. Other aids to good transmission are to adjust the microphone so that it is near but not too close to the mouth and to use a normal constant tone without speaking too slowly or too fast. Clarity of transmission and reception is important and the communications equipment should be checked by a **Radio Check** call. This is normally done by transmitting on the **Ground** or **Tower** frequency giving the following information:

Station being called	'Weston Tower'
Aircraft calling	'Golf Alpha Bravo Charlie Delta'
Purpose of transmission	'Radio check on'
Frequency	'One two zero decimal five'
Reply:	
Aircraft calling	'Golf Alpha Bravo Charlie Delta'
Station replying	'Weston Tower, readability four'.

This relates to levels of readability as follows:

Scale   1.   Unreadable
       2.   Readable now and then
       3.   Readable with difficulty

4. Readable
5. Perfectly readable

Identification of Air Traffic Services

A number of services are available to the pilot who will identify the service required using one of the following terms:

Control	Area control
Radar	Self explanatory
Precision	Radar for approach
Approach	Airfield approach control
Ground	Ground manoeuvring
Information	Flight information service within the FIR

Pilots approaching an airfield would call the ATS e.g. *Birmingham Approach* followed by the aircraft callsign and having established communication would then transmit the message.

Terminology

Certain words and phrases have been adopted as part of the general RTF phraseology; they should be memorized and clearly understood:

Radio Call	Meaning
Acknowledge	Confirm message understood.
Affirm	Yes.
Break	Indicates a clear separation of one part of the text and another or separate messages.
Cancel	Annul clearance transmitted.
Check	Examine system or procedure.
Cleared	Proceed with instructions.
Confirm	Re-advise.
Contact	Establish communication with–.
Correct	Self explanatory.
Correction	An error has been made, the correct information is–.
Disregard	Consider message not sent.
How do you read	Refers to Readability Scale.
I say again	I repeat for emphasis/clarity
Monitor	Listen out on (frequency).
Negative	No. Permission not granted/incorrect.
Over	Transmission ended but a response expected.

Out	Transmission ended no response expected.
Pass your message	Self explanatory.
Read back	Repeat back the message received.
Report*	Pass information required.
Request	I wish to know or do ...
Roger	Last transmission understood (Note **It is not the answer to a question**).
Say again	Repeat last transmission.
Speak slower	Self explanatory.
Standby	Wait until called.
Verify	Check and confirm.
Wilco	Instructions will be complied with.

*If an ATC instruction includes the word REPORT it may be assumed that the aircraft has been cleared to the position from where that report will be made. For example, in an airfield circuit an instruction to *report on final* indicates that the aircraft is cleared onto the final approach.

At the end of an ATC transmission the pilot must read back that part(s) concerning:

Altitude/Flight Level instructions;
Heading and speed instructions;
Airways and Route Clearances;
Clearance to enter, land; take off;
Back track or cross an active runway;
Altimeter Settings;
SSR (Transponder) Codes/Modes;
VDF (direction finding);
Frequency changes.

Priorities and Types of Transmissions

Aircraft flying on a great variety of operations and in all kinds of weather are subjected to ever-changing situations, some routine and others more exacting. Because radio messages from a number of aircraft cannot all be sent and received simultaneously a system of priorities exists.

Obviously a distress call must have priority over any other message or transmission. For this reason transmissions are divided into the following categories:

1. Distress
2. Urgency

3. Direction Finding
4. Flight Safety Messages
5. Meteorological Messages
6. Flight Regularity Messages

Flight safety messages are those which are transmitted by Air Traffic Control, or position reporting by aircraft. Meteorological messages are self-explanatory and flight regularity messages are related to the normal operation of aircraft, for example refuelling, servicing, alterations in schedules, etc.

VDF and Radar

VDF Direction Finding facilities make a useful contribution to navigation and a pilot may obtain headings to steer to or from the station using the Q Code.

QDM Magnetic heading to reach the station with zero wind.

QDR Reciprocal of QDM

QTE True bearing of aircraft relative to the VDF station.

As a general guide range is increased with height by 10 miles/1000 ft at the lower levels and it follows that if a position **Fix** is required i.e. a position obtained by the intersection of bearings plotted from two (or more) stations this could be difficult to achieve if the aircraft was below 3000 ft a.m.s.l.

Bearings are obtained by RTF transmission (usually the full aircraft callsign is sufficient) and are classified as follows:

Class A accurate to ± 2°
Class B accurate to ± 5°
Class C accurate to ± 10°

In the United Kingdom and some other countries there is comprehensive radar coverage which can give the aircraft's position almost instantaneously and with very great accuracy, a service clearly of immense navigational value to aircraft. Furthermore it is possible to obtain a highly accurate let down when the ground station is equipped with radar. Various types of equipment are in current use but in principle the ground controller informs the pilot of his relationship to the airfield or runway, giving him headings to steer (**Vector**) and descent instructions in such a way as to bring the aircraft within half a mile of the end of the runway, and on the centre line.

Emergency

One of the vital functions of RTF procedures is that information concerning flight safety can be passed to the appropriate ground station. A prime example of this is when an emergency makes it necessary for an aircraft to transmit a distress call.

Emergency calls may be considered under two headings, namely Distress and Urgency.

The distress call is prefixed by the words **Mayday-Mayday-Mayday**. This indicates that the aircraft is in imminent danger and requires immediate assistance.

Urgency communications (prefixed by **Pan-Pan-Pan**) are those which concern the safety of the aircraft, for example an undercarriage fault on a passenger aircraft would concern the safety of the aircraft but would not mean that it was in imminent danger.

The Distress Call

These are initiated and passed in this order:
'Mayday' spoken three times followed by the distress message, which should include the following information:

Callsign and aircraft type   *EIBUM   C152*
Nature of emergency   *ENG FAILU*
Pilot's intention e.g. ditching
Position
Altitude/Flight Level
Heading

If the aircraft is equipped with a transponder Code 7700 Mode C (Distress) should be selected to alert radar.

Having made this call on the frequency in use, if there is no response the pilot should change to the **International Distress Frequency** (121.5 Mc/s) and repeat the call. If for any reason the cause of the emergency no longer exists it would be necessary to cancel the distress call, an example being 'Mayday Golf Alpha Bravo Charlie Delta cancel distress, fire now extinguished, continuing to Manchester'. This call is acknowledged by the controlling ground station, who will, in order to resume normal communications, inform other aircraft that the distress traffic has ended. The aircraft will then revert to the original frequency on which the distress call was made and cancel the distress call on that frequency. The cancellation procedure is important since after the 'Mayday' call other aircraft must maintain **Radio Silence** until advised **Distress Traffic Ended**.

Urgency

The urgency message takes the same form as the distress message being prefixed by the words Pan-Pan-Pan. It takes priority over all messages except distress. It is of considerable assistance to the ATS unit involved to know the pilot's ability, e.g. student pilot, pilot with IMC rating, etc. as it allows a course of action to be planned within the capabilities of the pilot and the nature of the situation.

Assistance

At any time but particularly in the early stages of training perhaps on a solo cross-country flight a pilot may find he is having some difficulty with navigation or weather conditions and is reluctant to declare an emergency. The situation is recognized by ATC and a service known as Distress and Diversion (D & D) has been established within each FIR to provide the assistance required.

It must be emphasized that when a pilot recognizes that assistance is required, although no distress or urgency situation exists, he should not hesitate to inform ATC e.g. 'G-ABCD Cessna 152 student pilot uncertain of position in deteriorating weather'. These few words are all that is required to resolve the problem and prevent a situation from arising where a distress or urgency situation really does exist. The fact that you may be unsure of the correct RTF phraseology does not matter; what is important is to advise ATC.

One other word; ATC will pass some instructions, they may be related to frequency changes, altimeter settings, etc. When these are set, *double check for this is the time when mistakes often occur*.

Position Reporting

In flight under IFR liaison with ATC is an essential element in the conduct of the flight. The flight details, i.e. routeing, reporting points, flight levels, etc., will be known from the flight plan. Position reports will include the following information:

(*a*) Aircraft identification
(*b*) Position and time
(*c*) Altitude/Flight Level
(*d*) ETA next position.

In the case of flight outside controlled airspace in VMC the pilot may be advised to contact an appropriate ATS unit *en route*. Alternatively he may contact the FIR. London and Scottish Information is to some extent a dissemination service and as the flight progresses will where possible advise the pilot which next appropriate Air Traffic Service to contact. In this way ATC is aware of aircraft movements generally so contributing to the safety of air traffic flying VFR.

The frequencies for each service may be found in the COM section of the AIP and on a **Radio Facilies Chart** (a map of the country, or part of it showing the various ground stations and airfields with their related frequencies).

The required service is obtained by selecting the relevant frequency. The pilot is able to receive numerous facilities ranging from weather information to navigational assistance and let-downs.

Practical Examples of RTF

Situation

A student pilot is authorized to carry out a consolidation practice of circuits and landings, clearance has been obtained from ATC. After the appropriate aircraft checks aircraft calls:

Transmissions	Notes
A/c Birmingham Tower Golf Alpha Bravo Charlie Delta Radio Check one one eight decimal three.	Full callsign used until abbreviated by ATC.
Twr Golf Alpha Bravo Charlie Delta Readability five.	See Readability Scale (pages 325–6).
A/c Golf Alpha Bravo Charlie Delta taxi for circuit practice.	
Twr Golf Charlie Delta taxi to the holding point Delta Runway three three QNH one zero zero four QFE nine nine five.	Taxi clearances may specify a route. Movement is to a defined limit, in this case 'point Delta', on the airfield. Callsign abbreviated by ATC.
A/c Acknowledges and repeats altimeter settings. At the holding point power checks and vital actions are completed.	

A/c	Golf Charlie Delta ready for departure.	The words Take Off are used only for that specific event.
Twr	Golf Charlie Delta line up runway three three.	
A/c	Golf Charlie Delta lining up runway three three.	
Twr	Golf Charlie Delta cleared take off surface wind. three-two-zero, ten knots.	The circuit will normally be left hand, if not or if circuit directions are variable this will be advised.
A/c	Clear Take Off Golf Charlie Delta. (As soon as aircraft is position downwind.)	
A/c	Golf Charlie Delta downwind.	If other aircraft are in the circuit or in the approach phase ATC may allocate a priority No 1, 2, etc. A base leg call may be required.
Twr	Golf Charlie Delta report Final (approach).	
A/c	Golf Charlie Delta. (When turn onto final approach is complete.)	It is at this stage permission to land will be given. A position between 4 and 8 miles will be reported as long final and is appropriate to aircraft making a straight-in approach.
A/c	Charlie Delta final.	
Twr	Golf Charlie Delta Clear Land surface wind three one zero, ten knots.	In the case of circuit practice the aircraft may be cleared 'touch and go' i.e. touch down, roll and take off. This will have been advised in the earlier final call, viz. 'final for touch and go'.
A/c	Clear to land Golf Charlie Delta. OR	
Twr	Golf Charlie Delta Go around I say again Go around.	Should the pilot initiate the missed approach he will inform ATC.
A/c	Golf Charlie Delta Going around. At the end of the landing and as the aircraft slows on the runway.	
Twr	Golf Charlie Delta Vacate next left.	This instruction will be towards the end of the landing roll.

A/c Next left Golf Charlie Delta.

A/c Golf Charlie Delta runway vacated.

ATC may give specific taxi instructions where the situation is not obvious.

Twr Golf Charlie Delta taxi to Flying Club.

A/c Golf Charlie Delta Wilco.

Situation

A pilot wishes to fly from Whitecross, an airfield situated in a Control Zone, to Castleton 100 n.m. south-west. Flight conditions are VMC. He has booked out and is cleared to the holding point for runway 28.

Transmissions

Twr Golf Alpha Bravo Charlie Delta is clear to leave the Whitecross Zone on track Castleton maintaining VFR.

A/c Repeats the clearance.

Twr Golf Charlie Delta are you ready immediate departure.

A/c Golf Charlie Delta affirm.

Twr Golf Charlie Delta line up for immediate departure runway two eight.

A/c Golf Charlie Delta lining up runway two eight.

Twr Golf Charlie Delta cleared immediate take-off left turn out surface wind two seven zero, ten knots.

A/c Golf Charlie Delta Rolling, left turn out.

Twr Golf Charlie Delta airborne three zero [time] contact Approach one two two decimal five.

A/c one two two decimal five Golf Charlie Delta.

A/c Whitecross Approach Golf Alpha Bravo Charlie Delta climbing to Flight level four zero on track Castleton VFR.

Notes

ATC may impose height or heading restrictions until clear of the Zone.

App   Golf Charlie Delta Report
      Zone Boundary.

A/c   Golf Charlie Delta Wilco.

A/c   Golf Charlie Delta Zone
      Boundary.

App   Golf Charlie Delta you are
      cleared this frequency.
      (or Clear to QSY en route)

Alternatively there may be an
instruction to contact the next
appropriate service, e.g. LARS
(Lower Airspace Radar Service).
QSY – is a 'Q' code reference,
meaning to change frequency.

A/c   London Information Golf
      Alpha Bravo Charlie Delta.

In this case the pilot decides to
contact the FIR and pass his flight
details.

LI   Golf Alpha Bravo Charlie
      Delta London Information
      pass your message.

A/c   Golf Alpha Bravo Charlie
      Delta Cessna one five two
      from Whitecross to Castleton
      heading two two five
      Whitecross Zone Boundary
      three seven [time] FL four
      zero VFR estimating
      Castleton two nine.

Aircraft is flying a Quadrantal level
and above the transition altitude with
altimeter set to 1013.2 (Standard
Setting). Below the Transition
altitude he would use the Regional
QNH.

LI   Roger Golf Charlie Delta No
      known conflicting traffic
      contact Upton Radar one
      one nine decimal three.

To an extent the FIR is a
dissemination/co-ordination
service and will advise a next
appropriate service and in this case
a LARS south and abeam the route.

A/c   One one nine decimal three
      Good Day (more properly
      the callsign).

From time to time modest
departures from the formal RTF
phraseology do occur but excessive
courtesies are out of place.

A/c   Upton Radar –
      Message content to a LARS

      Full Callsign and aircraft type

      Estimated position

      Heading

      Altitude/flight level

      *Next reporting/turning
      point/destination.*

Rdr Golf Charlie Delta you are identified thirty miles north-east of Rockley maintain heading two three zero.

A/c Maintaining heading two three zero Golf Charlie Delta.

Aircraft receives a heading instruction which is repeated (page 327)

Rdr Golf Charlie Delta your position is now thirty miles north-east of Castleton heading good contact Castleton Approach one two zero decimal five.

A/c One two zero decimal five Golf Charlie Delta.

A/c Castleton Approach Golf Alpha Bravo Charlie Delta.

App Golf Charlie Delta we have your details; you are cleared to the field maintaining VFR Descend to two five zero zero ft QNH one zero zero nine.

A descent clearance for a VFR flight would not be given unless traffic considerations required. Flight details may be passed between ground stations.

A/c Leaving Flight Level four zero for two five zero zero ft on QNH one zero zero nine.

Changes standard altimeter setting at transition level for descent on airfield QNH.

A/c Golf Charlie Delta level at two five zero zero.

App Golf Charlie Delta Runway one three surface wind three one zero, ten knots. Visibility five km QFE one zero zero four report field in sight.

For aircraft below cloud in VMC and able to complete the landing in VMC, met. information can be limited to surface wind velocity and altimeter settings. Otherwise information given is:- Surface wind/visibility/present weather/cloud amount & base/ QFE, QNH/special information, e.g. RVR, gusting etc. At some airfields Automatic Terminal Information Service ATIS provides this pre-recorded information on a separate frequency. For the principal airfields it is also available on VOLMET broadcast half hourly, H+20 H+50

A/c Golf Charlie Delta Wilco
     QFE one zero zero four.

A/c Golf Charlie Delta field in
     sight.

App Golf Charlie Delta Contact
     Tower one one nine decimal
     five.

A/c One one nine decimal five
     Golf Charlie Delta.

A/c Golf Charlie Delta
     Downwind.

Twr Golf Charlie Delta Clear to
     Final.

A/c Golf Charlie Delta.

A/c Golf Charlie Delta Final.

Twr Golf Charlie Delta clear land
     surface wind three one zero,
     ten knots.

A/c Cleared to land Golf Charlie
     Delta.

Twr Golf Charlie Delta left at the
     intersection.

A/c Golf Charlie Delta.

A/c Golf Charlie Delta Runway
     vacated.

Taxi instructions complete the RTF
transmissions.

Aircraft could be cleared to base
leg or for a straight in approach
dependent on traffic. The presence
of other traffic will be advised.

Questions

1 A QDM is:
   (a) The magnetic bearing of an aircraft from a station,
   (b) The true bearing of an aircraft in relation to a station,
   (c) The magnetic heading to be steered by an aircraft to reach a
       station in conditions of no wind.

2 A class 'A' bearing is accurate within:
   (a) ±1°,
   (b) ±3°,
   (c) ±2°.

3 At lower levels the range at which satisfactory VHF bearings can be taken:
   (a) Increases by approximately 10 nm per 1000 ft additional height,
   (b) Is longer by day than at night,
   (c) Is shorter by day than at night.

4 **During a radio check with ATC the aircraft call is reported as 'Strength three'. The transmission is:**
   (a) Readable now and then,
   (b) Readable with difficulty,
   (c) Readable.

5 **High ground between an aircraft and a ground station can:**
   (a) Improve the range of VHF communications,
   (b) Interrupt VHF communications,
   (c) Distort VHF bearings without affecting communications.

6 **When an aircraft has established communications using the five-letter callsign it may subsequently be abbreviated:**
   (a) To the last two letters when two-way communications have been established,
   (b) To the last two letters when ATC first uses that abbreviation,
   (c) To the first and last two letters when ATC has initiated that call.

7 **Which of the following ATC messages requires 'readback' confirmation:**
   (a) Frequency change,
   (b) Take-off time,
   (c) Wind velocity on final approach.

8 **A student pilot is running out of fuel and in doubt of reaching the destination. The correct radio call would be to:**
   (a) Transmit a 'Mayday',
   (b) Send a Flight Safety Message,
   (c) Use the priority designated by the prefix 'PAN' repeated three times.

9 **After power checks and vital actions have been completed the pilot will:**
   (a) Request 'Take-off clearance',
   (b) Advise 'Ready for Departure',
   (c) Ask to 'Line up and Hold'.

10 **An aircraft flying VMC would, on establishing communications with Approach Control, be given:**
   (a) Descent and joining instructions,
   (b) Descent and landing instructions,
   (c) Airfield and Weather information.

11 **In an emergency that warrants a Distress Call the transponder (when fitted) is set to Code 7700 (mode C):**
   (a) To alert the radar service that a distress situation exists,
   (b) To warn other aircraft not to use their radio during distress communications,
   (c) To alert the airways controller that a distress situation exists.

12 **A Distress and Diversion service is provided:**
   (a) Within each of the Altimeter Setting Regions,
   (b) Within each of the Flight Information Regions,
   (c) Within the Lower Airspace Radar Advisory Service Area.

Appendix 1 — Answers to Multiple Choice Questions

Appendix I

Chapter 1. AVIATION LAW		Chapter 2. NAVIGATION		Chapter 3. METEOROLOGY	
1	(a)	1	(b)	1	(c)
2	(c)	2	(c)	2	(b)
3	(c)	3	(c)	3	(b)
4	(b)	4	(b)	4	(b)
5	(a)	5	(c)	5	(a)
6	(a)	6	(b)	6	(a)
7	(c)	7	(a)	7	(a)
8	(c)	8	(b)	8	(b)
9	(a)	9	(a)	9	(b)
10	(b)	10	(i) Spot Heights	10	(c)
11	(b)		(ii) Layer Tints	11	(c)
12	(c)		(iii) Contours	12	(a)
13	(c)		(iv) Form Lines	13	(b)
14	(b)		(v) Hill Shading	14	(a)
15	(a)		(vi) Hachures.	15	(b)
16	(c)	11	(c)	16	(b)
17	(c)	12	(b)	17	(c)
18	(b)	13	(c)	18	(b)
19	(c)	14	(a)	19	(a)
20	(a)	15	(a)	20	(c)
21	(c)	16	(a)	21	(b)
22	(b)	17	(c)	22	(a)
23	(a)	18	(a)	23	(a)
24	(b)	19	(b)	24	(b)
25	(c)	20	(c)	25	(a)
26	(c)	21	(a)	26	(b)
27	(a)	22	(b)	27	(b)
28	(c)	23	(c)	28	(b)
29	(c)	24	(a)	29	(a)
30	(a)	25	(b)	30	(b)
		26	(b)	31	(b)
		27	(c)	32	(c)
		28	(a)	33	(a)
		29	(c)	34	(b)
		30	(a)	35	(a)
		31	(b)		
		32	(c)		
		33	(a)		
		34	(b)		

Chapter 4. PRINCIPLES OF FLIGHT		Chapter 5. AIRFRAMES		Chapter 6. ENGINES AND PROPELLERS	
1	(c)	1	(b)	1	(c)
2	(a)	2	(a)	2	(b)
3	(a)	3	(c)	3	(a)
4	(a)	4	(a)	4	(a)
5	(c)	5	(c)	5	(c)
6	(c)	6	(a)	6	(b)
7	(a)	7	(b)	7	(a)
8	(b)	8	(a)	8	(c)
9	(a)	9	(c)	9	(a)
10	(c)	10	(c)	10	(a)
11	(c)	11	(c)	11	(b)
12	(c)	12	(b)	12	(c)
13	(b)			13	(a)
14	(a)			14	(c)
15	(a)			15	(b)
16	(b)			16	(c)
17	(c)			17	(a)
18	(a)			18	(c)
19	(a)			19	(c)
20	(b)			20	(b)
21	(c)			21	(a)
22	(a)			22	(b)
23	(c)				
24	(a)				
25	(b)				
26	(b)				
27	(b)				
28	(a)				
29	(c)				
30	(c)				
31	(a)				
32	(c)				
33	(c)				
34	(a)				
35	(b)				

Appendix I

Chapter 7. SYSTEMS		Chapter 8. AIRFIELD PERFORMANCE, WEIGHT AND BALANCE.		Chapter 9. INSTRUMENTS	
1	(b)	1	(a)	1	(a)
2	(b)	2	(b)	2	(b)
3	(a)	3	(c)	3	(c)
4	(a)	4	(b)	4	(a)
5	(a)	5	(c)	5	(a)
6	(b)	6	(a)	6	(c)
7	(a)	7	(b)	7	(b)
8	(c)	8	(c)	8	(a)
9	(b)	9	(c)	9	(c)
10	(b)	10	(a)	10	(b)
11	(c)	11	(c)	11	(b)
12	(c)	12	(c)	12	(a)
13	(a)	13	(a)	13	(a)
14	(b)	14	(c)	14	(b)
15	(c)	15	(c)	15	(c)
16	(b)			16	(a)

Chapter 10.
INTRODUCTION TO
RADIO TELEPHONY

1	(c)	17	(c)
2	(c)	18	(b)
3	(a)	19	(b)
4	(b)	20	(c)
5	(b)	21	(a)
6	(c)	22	(c)
7	(a)	23	(b)
8	(c)	24	(c)
9	(b)	25	(b)
10	(c)	26	(a)
11	(a)	27	(a)
12	(b)	28	(b)
		29	(b)
		30	(c)
		31	(c)
		32	(c)
		33	(b)
		34	(c)

Appendix 2 – Ground Signals

Runway or taxiway unfit for aircraft landing or manoeuvring.

Area to be used only for take-off and landing of helicopters.

Movement of aircraft and gliders confined to hard surface areas.

Direction of take-off and landing may differ. (Also shown by a black ball suspended from the signals mast.)

A red L on the above sign indicates that light aircraft may land on runway or grass area. (May also be used in conjunction with signal below.)

A yellow diagonal on a red square means special care is necessary in landing, owing to temporary obstruction or other reason.

Landing and take-off on runways only, but movement not confined to hard surfaces.

Landing prohibited — a yellow cross on a red square.

Glider flying in progress. (Used in conjunction with two red balls on the signals mast.)

Right hand circuit in force. The arrow is made up of red and yellow stripes.

Shaft of the T indicates direction of take-off or landing — i.e. towards the crossbar.

Part of the manoeuvring area to be used only for the take-off and landing of light aircraft.

Emergency landing only — two yellow bands on a red square.

Orange/white tent strips alternating with orange/white flags mark bad ground to be avoided while taxying. Airfield or other boundaries may be marked with the orange and white tent strips illustrated.

Yellow and red chequered flag (or board) denoting airfield control is operated by special signals. The letter C (on yellow background) marks pilot's reporting point. Two figures (also on yellow background) give direction of take-off and landing (magnetic) to nearest ten degrees. These figures are usually painted on the duty runway or in the case of grass airfields, on the landing and take-off strip in use at the time.

Appendix 3 — Explanatory Notes for LAPFORM 215

EXPLANATORY NOTES FOR LAPFORM 215 — CHART OF FORECAST WEATHER BELOW 15000 FT

1. **Purpose**

LAPFORM 215 is a forecast form which has been introduced to replace written low-level area and route forecasts for flights within the United Kingdom and to the very near Continent. The form covers in-flight conditions from the surface to 15000 FT and comprises three sections, a forecast chart, a tabular forecast, and relevant warnings and/or remarks. Both the chart and tabular section relate to the fixed time shown in the chart heading.

2. **Information given**

a. **Chart:**

   (i)   The forecast position, direction and speed of movement of surface fronts and pressure centres (see para 3 below).

   (ii) Areas of weather enclosed by continuous scalloped lines. Each area is allotted a distinguishing letter that cross refers to an entry in the tabular section on the right of the form.

b. **Tabular section:**

   (i)   The top line for each area gives the main expected conditions under the headings VIS (Visibility), WX (Weather) and TURBULENCE, CLOUDS, ICING, etc. If significant differences from the main conditions are likely within that area at the fixed time, these are given on subsequent lines after an explanatory remark in the VARIATION column.

   (ii) Surface visibility is expressed in metres using four figures up to 5000 M, and in whole kilometres using one or two figures for 6 KM or more.

   (iii) Cloud amount, type, and heights of base and top. All heights are expressed in hundred of feet AMSL. 'XXX' is normally used for tops in excess of 15000 FT but a height is always specified for CB tops.

   (iv) The expected occurrence of icing or turbulence is indicated using the standard symbols (see para 4 below). The forecast height range will be shown alongside the symbol unless the heights coincide with the full height range of the cloud with which they are associated. As indicated in Note 2 of LAPFORM 215 the words HILL FOG imply cloud covering hills, with a consequent visibility of 200 metres or less.

c. **Warnings and/or remarks box:**

   (i)   Significant changes expected (e.g. CB developing, fog dispersing) during the period stated in the top right hand corner of the form.

   (ii) Warnings to supplement the forecast information given in the tabular section immediately above (e.g. surface gales, low-level turbulence).

3. **Pressure centres and fronts**

The surface position of pressure centres is shown by X for a low centre and O for a high centre. The letter L or H, as appropriate, together with the central pressure in millibars are shown adjacent to the position.

The surface position of fronts is shown by the standard symbols, thus:

Warm front ●▲● Cold front ●▼● Occlusion ●▲▼●

The direction and speed of movement (in knots) of centres and fronts is given.
Movements of less than 5 knots are shown as 'SLOW'

Example:

H 1020 ● SLOW

L ×10↙ 992

4. Abbreviations and symbols used in the tabular section

Abbreviations:

SCT	– scattered (1/8–4/8)	used in relation to amount of cloud
BKN	– broken (5/8–7/8)	
OVC	– overcast (8/8)	
EMBD	– embedded (used only in respect of CB embedded in other cloud)	

ST	– stratus	CU	– cumulus	AC – altocumulus
SC	– stratocumulus	CB	– cumulonimbus	AS – altostratus
NS	– nimbostratus			

CAT	– clear air turbulence	COT	– at the coast	DZ – drizzle
FRQ	– frequent	ISOL	– isolated	LOC – locally
LYR	– layered	OCNL	– occasionally	TS – thunderstorm
WDSPR	– widespread			
SH	– showers (to be combined with type of precipitation)			

Symbols:
⇃ light airframe icing
⇂ moderate airframe icing
⇟ severe airframe icing

⩘ moderate turbulence
⩙ severe turbulence

5. Chart available at Fixed Time Validity Suitable for Initial Departures and Flights between (Period) (All times GMT)

0500	0800	0600 – 1100
0900	1200	1000 – 1500
1300	1600	1400 – 1900
1700	2100	1800 – 2400
2300	0300	0000 – 0600

NOTES

1. Where single numerical values are given for any element, these are to be interpreted as representing the most probable mean of a range of values covering approximately ± 25%.

2. Conditions quoted in the tabular section are for the fixed time given in the chart heading. Important changes with time will be referred to briefly in general terms in the Warnings and/or remarks box.

3. For detailed information on the expected changes at specific locations, e.g. destination airfields, the user should consult relevant TAFs or landing forecasts.

Index

(Figures in *italics* refer to illustrations)

Index

Index